RESOURCES OF REALISM

Resources of Realism

Prospects for 'Post-Analytic' Philosophy

Christopher Norris

Philosophy Section
School of English, Communications and Philosophy
University of Wales
College of Cardiff

First published in Great Britain 1997 by
MACMILLAN PRESS LTD
Houndmills, Basingstoke, Hampshire RG21 6XS and London
Companies and representatives throughout the world

This book is published in Macmillan's *Language, Discourse, Society* series
Editors: Stephen Heath, Colin MacCabe and Denise Riley

A catalogue record for this book is available from the British Library.

ISBN 0–333–67904–0 hardcover
ISBN 0–333–67905–9 paperback

First published in the United States of America 1997 by
ST. MARTIN'S PRESS, INC.,
Scholarly and Reference Division,
175 Fifth Avenue, New York, N.Y. 10010

ISBN 0–312–17551–5

Library of Congress Cataloging-in-Publication Data
Norris, Christopher.
Resources of realism : prospects for 'post-analytic' philosophy /
Christopher Norris.
p.   cm. — (Language, discourse, society series)
Includes bibliographical references and index.
ISBN 0–312–17551–5 (cloth)
1. Semantics (Philosophy)   2. Realism.   3. Empson, William, 1906–
. 4. Davidson, Donald, 1917–   . I. Title. II. Series.
B840.N67 1997
121'.68–dc21   97–6416
                                                            CIP

This book is printed on paper suitable for recycling and made from fully managed and
sustained forest sources.

10   9   8   7   6   5   4   3   2   1
06   05   04   03   02   01   00   99   98   97

Printed in Great Britain by
The Ipswich Book Company Ltd
Ipswich, Suffolk

For Frank Kermode

Contents

Acknowledgements

I would like to thank my friends and colleagues in the Philosophy Section at Cardiff for their interest in this project and their many acts of kindness, support, and encouragement as the work went along. To Michael Durrant I am especially grateful for hearing me out on numerous occasions and suggesting some much-needed technical refinements. It was my great good forture (if not perhaps his) that we occupied adjoining rooms while this book was nearing completion. The idea of writing it first came from an open debate on the topic of 'Truth' with Peter Sedgwick and Alessandra Tanesini. If we are not much closer to agreement now – some four years on – it has none the less clarified my thinking on several points and provided plentiful matter for discussion in and out of the seminarroom. I am also much in debt to my past and continuing PhD students, especially to Christa Knellwolf who read an early version of the typescript and suggested many improvements of structure and detail. Clive Cazeaux, Gideon Calder, Carol Jones, Marianna Papastephanou, Daniel Procida, and David Roden have been writing on topics most closely related to the interests of this book and should therefore take a part of the credit (though none of the blame) for the way it has turned out. Wendy Lewis produced a splendid piece of artwork for the jacket and once again proved me wrong in my decided belief that the book contained nothing remotely conducive to visual inspiration. Joe Margolis's work has been a constant inspiration, so I trust he won't object to the adversary role thrust upon him in Chapter 6. To Alison, Clare and Jenny – yet again – my thanks for just about everything.

Introduction

This book turns a critical (though not I hope a jaundiced) eye on various ideas and movements of thought that are nowadays grouped under the broad rubric of 'post-analytic' philosophy. What chiefly unites them – on the negative side – is a growing dissatisfaction with the analytic enterprise as it developed in the wake of logical empiricism. That project is now taken to have failed in all its main objectives, among them more recently the attempt to develop a truth-theoretic compositional semantics for natural language and a theory of beliefs (or propositional attitudes) that would explain how speakers and interpreters display such remarkable – though everyday – powers of communicative grasp. These ideas have come under attack from many quarters during the past two decades. Most influential here has been Quine's assault on the two 'last dogmas' of empiricism and – supposedly following from that – his case for ontological relativity and meaning-holism as the only way forward in default of any method for individuating objects or items of belief. The result, very often, is an attitude of deep-laid scepticism with regard to the truth-claims of science and the idea that philosophy might offer grounds – reasoned or explanatory grounds – for our understanding of language and the world.

Hence the turn toward different conceptions of philosophical enquiry which renounce those false 'objectivist' ideals and instead take their bearing from various alternative sources. Among them – most recently – are narrative pragmatics, hermeneutics, Heideggerian depth-ontology and the familiar late-Wittgensteinian appeal to language-games (or cultural 'forms of life') as the end-point of all such questioning. These approaches are often combined with some version of the 'strong' programme in sociology of knowledge wherein science is conceived as just one (currently prestigious) language-game devoid of any special epistemic or explanatory warrant. This case has found support from various, otherwise disparate philosophical positions. On the one hand are those – like Richard Rorty – who welcome the collapse of that old 'analytic' paradigm as a sign that philosophy has at last relinquished its grandiose epistemological pretensions and accepted its role as just one (strictly non-privileged) participant voice in the ongoing cultural 'conversation of mankind'. On the other are thinkers such as Donald Davidson who have striven to defend some version of a truth-based semantic theory but failed to specify its structure and content in adequately detailed or substantive terms.

Thus in Davidson's view our last, best hope of countering the sceptical-relativist trend is to invoke a Tarskian 'disquotational' theory of truth which reduces to a purely formal specification ('"snow is white" is true if and only if snow is white'). Such a theory automatically applies by definition to each and every true sentence in each and every language. However, it can plausibly be argued that 'true' here functions as the merest of place-holding predicates, that is to say, as a term that denotes no more than the tautologous (redundant or circular) equivalence between the two sides of a Tarskian T-sentence. In which case the theory amounts to just a stopgap means of preserving some technical notion of truth which can always be absorbed into the equation by the simple expedient of cancelling through on both sides and thus producing a straightforward statement of empirical fact. By the same token it offers absolutely no help in explaining just *why* – in virtue of what real-world facts, properties, causal attributes, belief-ascriptions, and so forth – we might be warranted in attributing truth or determinate content to certain propositions and falsehood (or present undecidability) to others.

Hence the change of heart among some philosophers – notably Stephen Schiffer – who started out on the quest for a truth-based intentionalist semantics but who now think that enterprise hopelessly stalled on problems of meaning-variance, ontological relativity, the range of diverse (incompatible) belief-attributions, and so forth. In philosophy of science also there is a widespread acceptance of the Kuhnian thesis that any 'truths' to be had are truths only relative to this or that paradigm or framework of knowledge-constitutive norms, taken to define the very 'world' – or horizon of intelligibility – within which scientists (and philosophers or historians of science) propose their various, strictly 'incommensurable' claims. Nor is it much use invoking the Davidsonian 'Principle of Charity' as a weak transcendental argument to the effect that we and other people (other language-users, cultural communities, scientific worldviews, etc.) *must* be right about most things since the notion of massive systematic error is a simply unintelligible thesis which nobody (we or they) could be in a position to advance or ascertain. For it is, to say the least, a very

dubious form of *a priori* argument that brings people out necessarily 'right in most matters' and which takes no account of the various factors – causal factors among them – which may help to
explain why they (or we) had been led to form erroneous beliefs or
to misinterpret the beliefs of others.

Thus 'charity', as Davidson construes it, amounts to just a kind of
enabling fiction adopted in order to maximize truth-ascriptions –
and thereby preserve the possibility of mutual understanding –
irrespective of whether the meanings, intentions, beliefs, or truth-
claims in question have any genuine descriptive, explanatory, or
veridical warrant. Hence Davidson's subsequent recourse to a
minimalist semantics which finds little use for 'prior theories' of the
sort that might go some way toward explaining just how it is that
language-users do in fact manage to communicate across sizeable
differences of context and cultural background. Rather, we should
rely on 'passing theories' whose field of application (as Davidson
puts it) may be 'vanishingly small', and which really come down to
mere 'wit, luck, and wisdom' plus a readiness at times to disregard
the lexico-grammatical sense of what people say – as given by a
prior theory – in favour of guessing what they most likely mean in
this or that context of utterance.

So it is that Davidson can assert 'there is no such thing as a
language' if by 'language' is meant the sort of thing that linguists,
philosophers, and (it might be thought) everyday language-users
have in mind when they reflect on what is involved in the process
of communicative uptake. For there now seems to him little hope of
redeeming any project of 'radical translation' – or even any theory
of how speakers and interpreters communicate under fairly normal
circumstances – given the problems that arise for such a theory in
the wake of Quinean and related arguments in the post-analytic
(ontological-relativist) mode. That is to say: if we lack any reliable
means of individuating beliefs (or utterer's intentions) through the
analysis of truth-conditions, semantic structures, or propositional
content then neither can we make much sense of the idea of 'possessing' or 'sharing' a language. That idea could only have explanatory value if it captured something intrinsic to the nature of
linguistic understanding in general. But this cannot be the case – so
Davidson now thinks – since we can always carry on talking and
figuring out each others' beliefs, meanings and intentions (at least
for most practical purposes) despite the impossibility of providing
any such generalized theory.

In this book I take a different, more sanguine view of the prospects for redeeming that supposedly forlorn enterprise. These problems have arisen, I suggest, chiefly as a result of the logical-empiricist programme which conceived of truth as a formalized relation between first-order statements of empirical fact and a second-order (metalinguistic) discourse of generalized validity-conditions. It is this particular strain of analytic thought – along with its various successor-movements – that has narrowed the options in recent debate, thus provoking a full-scale movement of retreat among erstwhile adherents like Schiffer and a quest for alternative options among born-again pragmatists such as Rorty and half-way converts to the pragmatist cause like Davidson and the later Putnam. In philosophy of science, likewise, the choice has too often seemed to fall between a logical-empiricist approach whose failings have long been apparent and a turn toward other – depth-hermeneutic or 'strong' sociological – paradigms which count all talk of truth and reality a world well lost for the sake of such new-found interpretive freedoms.

My point, in each case, is that these are demonstrably false dilemmas whose effect has very often been to close off other, more adequate and promising avenues of philosophical enquiry. In the chapters that follow I aim to provide an extended and integrated treatment of several such alternative projects. They include the causal-realist approach to issues in philosophy of science developed by theorists such as Wesley Salmon and Roy Bhaskar; the causal theory of reference, naming and necessity (construed in opposition to various forms of descriptivist or conventionalist doctrine); and a naturalized (but non-reductionist) theory of language and interpretation whereby to secure substantive content for various beliefs (true and false) that can then be subject to rational assessment on a basis of potentially shared understanding. For there is, I shall argue, nothing incompatible about causal theories of knowledge-acquisition and epistemic theories that require some further (realist or verification-transcendent) criterion of truth and falsehood. These approaches will only seem mutually exclusive if one takes it – like Hume and his latter-day descendants – that there exists a methodological gulf between causally operative processes and the logic of scientific or philosophical enquiry.

What is needed, therefore, is a truth-theoretic intentionalist semantics which rejects that artificial dichotomy and locates the structure and content of truth in forms of natural-language expression. Just such a theory is to be found in William Empson's book *The Structure of Complex Words,* a work that has been pretty much ignored by philosophers but which offers by far the most sustained and resourceful treatment of these issues in philosophy of language and interpretation-theory. More specifically it shows – in contrast to Davidson's minimalist-semantic approach – how meanings, attitudes, and beliefs can be analysed in terms of propositional content (or structures of logico-semantic entailment) which then give a hold for construing speakers' intentions. Also Empson offers some cogent criticism of holistic or contextualist theories which effectively relativize meaning and truth to this or that currency of in-place belief or interpretive 'horizon' of enquiry. For in that case – I argue – there could be no explaining how speakers and interpreters manage to exhibit such otherwise preternatural powers of linguistic creativity and grasp.

Just lately there has emerged a curious *rapprochement* between post-analytic philosophy and a selective neo-pragmatist reading of Heidegger that accommodates certain of his depth-ontological or hermeneutic themes. For some commentators – Mark Okrent, Hubert Dreyfus, and Richard Rorty among them – this points a way forward from the dilemmas of logical empiricism and the formalized Tarskian theory of truth with its (arguably) trivial or redundant outcome. However – I shall urge – there is little to be gained from an alliance with those aspects of Heidegger's thought (in particular concerning language and the so-called 'question of technology') that leave no room for any viable account of human intentions and ethical responsibility. More promising by far is that other kind of depth-ontological enquiry which respects the achievements of the physical sciences and takes them as the basis for a realist and causal-explanatory theory of knowledge acquisition. What counts as explanatory 'depth' in this context is a grasp of those underlying properties – causal dispositions, microstructural attributes, subatomic configurations, etc. – which provide a more adequate and detailed understanding of objects or events in the physical domain.

It is here that philosophy of language has most to learn from developments in recent neo-realist epistemology and philosophy of science. Hence the arguments put forward by Saul Kripke and (at one time) Hilary Putnam for a causal theory of naming and necessity construed by analogy with discovery-procedures in the natural sciences. In brief, there exists an order of *a posteriori* (empirical) necessity which applies both to items of scientific knowledge – like the definition of 'water' as that substance with the molecular composition H_2O or 'gold' as that element with atomic number 79 – and also to the way that such names are secured as referring to just those kinds of substance in virtue of the knowledge thus acquired concerning their salient properties and features. The need for some such account is clearly shown by the difficulties that Quine runs into when he endorses the claim of the physical sciences to be our best, most reliable source of knowledge, despite and against his own express commitment to a doctrine of wholesale ontological relativity. On the alternative (Kripke/Putnam) view these problems arise only at the point where descriptivist theories of meaning and interpretation lose touch with the various empirically tested and conceptually well-defined discovery procedures that characterize the advancement of scientific knowledge.

In short, there exist good arguments against the strain of anti-realist thinking that has gained wide acceptance in recent philosophical debate. (It is also pretty much *de rigueur* for a diverse company of post-structuralists, postmodernists, and cultural or literary theorists working with a thoroughly inadequate theory of language and representation derived from a selective reading of Saussure on the 'arbitrary' nature of the sign.) Of course there are certain aspects of recent speculative science that appear to lend support to a global anti-realist or purely instrumentalist stance. In one field especially – that of quantum mechanics – there might seem little choice but to adopt an interpretivist viewpoint that treats any truth-claims or observation-data as always subject to the kind of ultimate undecidability that results from the limits of precise measurement or the lack of objective criteria for assigning definite values to statements of, for example, particle location and momentum. As against this view I argue that ontological realism is *presupposed* by any thought-experiment or speculative theory – such as those advanced by Einstein and Bohr in their famous series of debates – that would carry some measure of demonstrative or probative force. For otherwise such theories would be wholly on a par with science-fiction fantasy or the wilder sorts of counterfactual imagining. That is to say, they would lack a referential bearing (or a specified object-domain) in respect of which their various truth-claims, hypotheses, or conjectures could be open to rational assessment and criticism. This applies just as much to quantum-theoretical conjectures as to classic thought-experiments – like those of Galileo – whose validity-conditions are a matter of jointly logico-deductive and empirical warrant. These issues have also loomed large in recent debate on the scope and limits of human understanding in general. Thus for Quine one argument that counts strongly in favour of ontological relativity is the doubt introduced by quantum-mechanical considerations when physicists (or philosophers) attempt to decide between rival ontological schemes. For Putnam likewise the retreat from a strong causal to a weak ('internal') realism appears to have been prompted in part by reflection on the status of scientific truth-claims *vis-à-vis* the need for a heterodox 'quantum logic' that would abandon certain well-entrenched classical axioms such as non-contradiction and excluded middle. Thomas Kuhn's ideas about radical para-digm-incommensurability can also be seen as a backward projec-tion from the current (exceptionally prolonged) 'revolutionary' situation in theoretical physics to the previous history of scientific methods and endeavours.

Elsewhere – in the case of postmodernists like Lyotard – it is pretty much taken as read that these quantum-mechanical 'para-doxes' (along with an assortment of kindred *topoi*: complemen-tarity, chaos-theory, fractals, Gödel's undecidability-theorem, etc.) are signs that knowledge has now moved on into a phase where 'performativity' (rather than truth) is the operative name of the game. Thus science should strive to maximize 'dissensus' and pro-duce the greatest range of heterogeneous narratives, discourses, or 'phrase-genres', rather than fixing its sights on some delusory idea of rational consensus or truth at the end of enquiry. It is, I think, important to see what is wrong with these modish variations on the ontological-relativist theme since they involve not only a loose extrapolation from vaguely grasped scientific theories but also some deeply obscurantist ideas about the ethical and socio-political bearings of (so-called) 'postmodern' science.

In short, I argue the case for a realist ontology, a truth-functional semantics and theory of belief-content, and a naturalized epistem-ology that none the less allows – indeed requires – the further appeal to context-transcendent criteria of truth and falsehood. This in turn means rejecting the narrowly extensionalist (Quinean) approach to these issues which would have it that epistemology can be naturalized only on condition that it exclude all talk of meanings, intentions, beliefs or propositional attitudes. The latter I take to be a wholly unwarranted requirement given the alternative resources outlined above and the manifest shortcomings of the Quinean doctrine when applied not only in the context of natural-language understanding but also to issues in epistemology and philosophy of science. For here also – as Roy Bhaskar points out – there is a need to distinguish the domain of 'intransitive' (real-world and causally efficacious) objects, processes and events from the 'transitive' realm of humanly intelligible purposes, motives, and intentions. Only by respecting this cardinal distinction can philosophy preserve a due sense of the limits – but also the significant scope – for an ethically and socially responsible critique of science and its human consequences. Otherwise it is prone to the twin fallacies of a hardline reductionist physicalism on the one hand and, on the other, a Rorty-style notional freedom to redescribe reality however we please, unconstrained by any merely 'ontic' concern with those features of the physically existent world – causal regularities among them – that might fail to accord with such descriptions.

The same kinds of argument may be seen to apply in philosophical semantics and other disciplines (cognitive psychology among them) having to do with matters of truth, meaning, and belief-ascription. For counting other people right in the main– as required by the Davidsonian principle of charity – must always be balanced against the standing possibility that they might yet be wrong (understandably wrong) with regard to certain items of belief arrived at on the basis of false information, as a result of some perceptual anomaly, or through various factors that can best be reckoned with in causal-explanatory terms. Hence the alternative 'principle of humanity' (a term I borrow from Richard Grandy and David Papineau) which makes it a rule to maximize truth-content while none the less allowing for just such sources of humanly intelligible error. Here again what is needed is a naturalized epistemology which takes due account of these factors but which also preserves the crucial (transcendentally justified) distinction between values of truth and falsehood.

All of which amounts to a strong case – I would argue – for rejecting the current anti-realist trend in epistemology, philosophical semantics, and interpretation-theory. For the way is then open to explore a wide range of developed (though as yet uncoordinated) lines of approach that pay more regard to the practical contexts of linguistic grasp and knowledge acquisition.

(Note: I have not provided references for this introductory essay since the various authors and topics here mentioned are all discussed at greater length – and with full bibliographical details – in subsequent chapters. The reader will be able to locate these details readily enough by cross-referring *via* the index of names and relevant endnotes.)

1

Minimalist Semantics and the Hermeneutic Turn: On 'Post-Analytical' Philosophy

AN ODD COUPLE : DAVIDSON AND HEIDEGGER

In recent years there have been various, more or less resourceful attempts to establish an alliance between so-called 'post-analytic' philosophy and hermeneutics in the Heideggerian or depth-ontological mode. Richard Rorty has made a number of essay-length overtures in this direction, though his promised book on Heidegger has never materialized, perhaps on account of certain difficulties which I intend to examine here.[1] Other writers – Mark Okrent among them – have all the same pressed ahead with this project to the point of claiming Heidegger as a kind of honorary pragmatist *malgré lui*, one who can be coaxed back into the fold by stressing his talk of situated being-in-the-world and tactfully downplaying his other, more *echt*-ontological themes and concerns.[2] Hubert Dreyfus's commentary on *Being and Time* accords them a somewhat more respectful hearing but still stresses their practical or everyday usefulness as a means of just finding our way about the world and dispelling certain needless metaphysical illusions.[3]

For the moment I shall steer clear of these Heideggerian depths and look more closely at the current situation in 'post-analytic' philosophy. We may then be better placed to understand just why there has emerged this curious *entente cordiale* between hitherto sharply divergent (indeed antagonistic) schools of thought. For it was, after all, a main plank in the programme of analytic philosophy – announced most clearly in Carnap's essay 'The Elimination of Metaphysics through Logical Analysis of Language' – that Heideggerian depth-ontological talk should be counted just a species of bewitchment by language, a kind of *ersatz* poetry which

1

(unlike the real kind) took itself to possess genuine veridical con-tent.[4] No doubt one reason for this latter-day *rapprochement* is the extent to which logical empiricism of the Carnapian variety has come under attack from various quarters during the past three decades and more. Most influential here has been Quine's 'Two Dogmas of Empiricism', where he rejects not only the distinction between analytic and synthetic propositions, but also the poss-ibility of drawing any firm or categorical line between observation-statements, logical 'truths of reason', and other (as Carnap would have had it) 'meaningless' or merely 'metaphysical' sorts of ex-pression.[5] All of which may help to explain the lowered resistance to Heidegger – and to 'Continental philosophy' in general – among thinkers of a post-analytic persuasion. For it now seems to many that the logical-empiricist programme was simply the first, most extreme and doctrinaire version of a project that has run aground on difficulties of its own artificial contrivance. In which case per-haps the way is open to a fruitful exchange with that whole 'other' tradition whose chief representatives – the post-Kantian line from Hegel, *via* Schopenhauer and Nietzsche, to Husserl, Heidegger, Gadamer and even Derrida – may yet have something useful to contribute.

What these approaches share is a sense that philosophy has now arrived at a stage – with its holistic turn against any version of the logical-empiricist paradigm – where talk of 'truth' (as hitherto con-ceived) becomes pretty much redundant. That is to say, it either drops out altogether (as in Rorty's neopragmatist appeal to what is 'good in the way of belief'),or else figures merely as a product of formal definition.[6] Such is the 'disquotational' theory – devised by Alfred Tarski and taken up by Donald Davidson – where 'true' has the role of a metalinguistic predicate that applies to each and every veridical statement in a given language, but which then cancels out – for all practical purposes – so as to leave those first-order state-ments quite unaltered.[7] Thus for instance: '"snow is white" is true in language L iff [i.e., if and only if] snow is white'. By means of this formal notation – so the argument runs – one can produce a corre-sponding 'T-sentence' for every sentence in the object-language. And one can then go on – by a simple recursive procedure – to generate a full-scale theory for that language which matches those sentences one by one in respect of their meaning (or truth-value) as given by the standard schema.

However, the problem with this line of thought is that it offers nothing more than a purely circular definition of 'truth', one that satisfies the formal requirements for such a theory while failing to provide any more specific or substantive set of criteria. As J.E. Malpas remarks, 'we ordinarily think about truth as a matter of sentences being true in virtue of the way things are in the world'. But the Tarskian theory responds to this requirement 'only in a technical and somewhat attenuated sense'. That is to say, it provides not so much a definition of 'truth as such', but only 'a definition of the applicability of the truth predicate for a particular language'.[8] Thus one can specify the formal truth-conditions for each sentence in the first-order (object) language by applying the standard Tarskian procedure which pairs it with a second-order (metalinguistic) counterpart sentence whose terms, logical structure, and extension are of course precisely equivalent. This produces a construction (a 'T-sentence') of the form: 'The sentence s of language L is true if and only if p', where s is the left-hand sentence as quoted (or 'mentioned') in the metalinguistic mode, while p – occurring on the right-hand side – is the identical sentence as straightforwardly deployed in an act of assertoric utterance. Whence, to repeat: ' "snow is white" is true in L if and only if snow is white'.

Such a theory, Malpas writes, 'will not, of course, provide a definition of truth *simpliciter*, but a definition of truth-in-L, that is, a definition of truth as it applies in the object language'.[9] It is 'disquotational' in the sense that one can then – as Davidson is wont to say – re-establish unmediated contact with the world by lifting the metalinguistic markers (i.e., the left-hand double quote marks around "snow is white"), erasing the right-hand sentence altogether, and thus coming out – after all these complex operations – with a plain statement of fact ('snow is white'), now restored from the context of 'mention' to that of 'use'. Some philosophers, Davidson among them, have made large claims for this theory. On their account it offers the best – perhaps the only – way forward from the doctrine of radical meaning-variance across and between language-games, paradigms, discourses, interpretive frameworks, and so on.[10] For the upshot of that doctrine – as in the work of thinkers such as W.V. Quine and Thomas Kuhn – is to generate an outlook of wholesale ontological relativity and, coupled with that, an attitude of deep-laid scepticism as regards the prospect of interlinguistic or transcultural understanding.[11]

Such is Davidson's argument in his well-known essay 'On the Very Idea of a Conceptual Scheme'.[12] His recommendation, in brief, is that we give up thinking of 'truth' as relative to (or 'constructed in') this or that language, ontology, conceptual framework, structure of semantic presuppositions, or whatever. For this is to get the matter backward, according to Davidson. Rather, it is the notion of truth – or the attitude of holding-true – that must be taken as basic to all language and which therefore provides at least a minimal starting-point for the process of figuring out what speakers mean in otherwise (to us) quite opaque contexts of utterance. Davidson's chief target here is the Quinean idea of 'radical translation', that is, the famous thought-experiment in which an anthropologist attempts to compile a translation-manual for some remote or hitherto unknown language and culture.[13] Quine's point is that even with the best-willed 'native informant' – and even where the context might seem wholly unambiguous, as for instance if he or she gestured toward a rabbit and produced the utterance 'gavagai!' – still the anthropologist couldn't be sure that 'gavagai = rabbit' as a matter of straightforward definitional equivalence. For the informant might have been saying (or intending) a whole range of other, less directly informative things, such as 'nice fluffy creature', 'good to eat', 'saw one like it yesterday', or (Quine's own exotic specimens) 'undetached rabbit-part' or 'temporal slice of rabbithood'. In other words there is no *a priori* reason to suppose that the act of ostensive definition (i.e., pointing at an object and uttering its name) either functions in the same way from one culture to the next, or provides any sure criterion for picking out intended objects of reference.

Davidson's response to all this is quite simple. In order to locate such possible breakdowns in communication we have to start out from the basic assumption that the native informant must hold certain things true. From which it follows that his or her language must possess at least some basic apparatus for handling matters of reference, predication, logical inference, evidential warrant, and so forth. This is why syntax, in Davidson's laconic phrase, is so much more 'social' than semantics. For it is a shared feature of various present-day relativist doctrines – whether Quinean, Kuhnian, Foucauldian, post-structuralist, or inspired by the ethno-linguistic speculations of a thinker like Benjamin Lee Whorf – that they all move straight from a semantics-based conception of language to a doctrine of full-scale meaning-holism which is taken to exclude (or radically to problematize) the possibility of translating with any degree of assurance between one and another language, discourse, paradigm or conceptual scheme. But the picture changes decisively, so Davidson would urge, when we switch attention to the 'syntax' – to the logical connectives, quantifiers, predicative functions or anaphoric devices – in the absence of which no language could achieve a level of functional adequacy. What then becomes clear is that sceptics or relativists like Quine, Foucault, Whorf and company are tacitly relying on those same interlingual resources even as they seek to conjure up the spectre of 'radical translation' as a strictly impossible enterprise.

Thus Whorf, while arguing that Hopi Indian and English cannot be 'calibrated', still purports to offer an English translation of sample Hopi sentences and – what's more – to describe some of the salient differences between their mental universe (ontology or worldview) and our own.[14] There is a similar problem with Kuhnian talk about the radical 'incommensurability' of scientific paradigms. For in fact Kuhn manages to explain quite convincingly – even with respect to 'revolutionary' periods of crisis and paradigm-change – how these shifts came about and what effect they had on the operative meaning of terms like 'mass', 'gravity', or 'combustion'.[15] In Quine's case likewise it is hard to reconcile his express doctrines of wholesale meaning-variance and ontological relativity with his professed high regard for the physical sciences as our best, most reliable means of enquiry and his forthright commitment to various theses regarding the scope and limits of knowledge in general.[16] Such – briefly stated – is Davidson's argument for rejecting any version of meaning-holism that relativizes 'truth' to some particular language-game, paradigm, ontology, semantic framework, or conceptual scheme. Thus philosophers are wrong – 'get the matter backward' – if they treat convention (defined in these various ways) as the precondition for language, and language in turn as the precondition for whatever counts as 'true' by the lights of some particular (cultural or interpretive) community. On the contrary, Davidson asserts: truth (or the attitude of holding-true) is a logically primitive notion, one that is presupposed in every act of understanding, whether within or between such communities. In his view the Tarskian schema provides the best means of rendering this argument in a mode that possesses both formal adequacy – since it extends, by definition, to every true sentence of the object-language – and also a strong measure of intuitive self-evidence.

Nevertheless there is clearly a sense (as Malpas remarks) in which a theory of truth along these lines does nothing whatsoever to explicate that notion in other than purely tautological or circular terms. Davidson himself seems prone to doubt on this question. In 'The Very Idea of a Conceptual Scheme' he takes it that the Tarskian formula is sufficient to define 'truth' in all relevant contexts, or at least to dispel the sorts of hyperinduced problem that typically arise when philosophers, linguists, cultural theorists and others reverse the logical order of priority between truth and meaning. But elsewhere – as for instance in his later series of lectures on 'The Structure and Content of Truth' – Davidson appears less sure of his ground.[17] In fact he concedes at the outset that the Tarskian schema is a purely formal device, one that of necessity (in virtue of its logical structure) defines 'truth' for each and every sentence of the object-language, but which offers nothing more substantive by way of explaining just what such truth might amount to in any given case. Nor has the position been much clarified by the close of Davidson's lecture-series. Thus he starts by suggesting (as indeed did Tarski) that some further notion of truth-as-correspondence may be required in order to give the theory content and save it from empty circularity.[18] But this promise remains unfulfilled since Davidson fails to specify that content with any degree of precision. In the end he falls back upon a psychologistic account of the attitude of 'holding-true', that is, something more like a coherence-theory where true statements are those that hang together with the whole current set of belief-dispositions to which a given speaker might signal his or her assent.

So one can see why many commentators – Rorty, Dreyfus, Okrent, and Malpas among them – have sought a way beyond what they perceive as this dead-end predicament. Such is at least one sense of the term 'post-analytic philosophy': the quest for an alternative to that entire tradition of thought, starting out from logical empiricism, whose upshot – after so much critical labour – would seem to be *either* a formalized (semantic or metalinguistic) theory of truth devoid of explanatory content, *or* on the other hand a pragmatist conception that reduces truth to the currency of in-place consensus belief. For each of the above thinkers it is clear that this alternative must come from outside the analytical mainstream, and moreover, that it must involve something in the nature of a Heideggerian (or depth-hermeneutic) approach to questions of meaning and truth. As I have said, they differ quite considerably in the extent of their commitment to Heidegger's project, that is, in their willingness to value his thought at its own 'epochal' or world-transformative estimate, rather than treating it – like Rorty – as just a source of new language-games, 'final vocabularies', or 'metaphors we can live by'. Indeed one could argue that it is a mark of the post-analytic appropriation of Heidegger – as distinct from the work of *echt*-Heideggerian disciples – that the former sorts of commentary always involve some degree of doctrinal non-attachment, some reluctance to endorse his more portentous 'ontological' pronouncements. Rorty is of course the most explicit about this, since for him what is of value in Heidegger (as likewise in Hegel, Nietzsche, Wittgenstein, Foucault, Derrida *et al.*) is simply his offering a novel set of terms for the ongoing 'cultural conversation', and thereby helping to wean us away from old (e.g. Kantian or 'analytic') styles of talk. Dreyfus and Okrent are perhaps best seen as half-way converts with a strong pragmatist leaning and with at least sufficient analytical awareness to resist the Heideggerian 'jargon of authenticity' in its more potent (or virulent) strain.

Malpas presents a rather different case. His book (*Donald Davidson and the Mirror of Meaning*) works patiently through all the above-mentioned problems with Davidson's thinking and then proposes the turn toward Heidegger as a means of rescuing truth from its reductive treatment as a place-filling (formalized semantic) substitute for the logical-empiricist doctrine. The following passage captures all the main details of his argument so I shall quote it *in extenso* to avoid a much lengthier paraphrase.

> Recognition of the need to conceive of truth in a more fundamental way arises most clearly out of consideration of traditional theories of truth in the light of Davidsonian holism. We have already seen the impossibility of providing an account of the notion of truth which does not itself rely on some prior notion of being true. Correspondence, coherence, even pragmatism, can provide no account of truth as such. Davidson's use of Tarski, while it is often taken to embody a correspondence account of truth, itself relies on a notion of truth, rather than providing any fundamental explication of the notion. The Tarskian account relies on interpreters' prior grasp of the concept of truth as it is embedded in their understanding of their own language. That grasp of truth as it operates in language is, moreover, itself embedded in our understanding of the world. Thus Davidson claims

that knowing a language – knowing a truth-theory – cannot be distinguished from knowing our way around the world in general. An understanding of truth is also, therefore, an understanding of the world. If this seems a strange thing to say, it nevertheless makes good sense in Heideggerian terms. For the understanding involved here can be seen as really that pre-understanding that is the original opening or unconcealment of things which is the ground of our experience and understanding of objects.19

There are similar arguments to be found in Dreyfus, Okrent and Rorty, each to the effect that analytical philosophy has in some sense arrived at a terminal impasse. Thus the choice falls out between a formalized redundancy-theory of truth and – what amounts to much the same thing on this reading of Davidson – the pragmatist idea that logic, language and truth come down to just a matter of 'knowing one's way about in the world'. And indeed there are passages in Davidson where he seems quite happy to endorse such a reading. Thus, '[o]f course truth of sentences remains relative to language, but that is as objective as can be'.20 For these problems will disappear at a stroke – so Davidson suggests – if one simply takes the view that 'knowing a truth-theory' amounts to nothing more than figuring out beliefs on an *ad hoc* basis of plausible guesswork from one situation to the next. And this in turn requires only a 'passing theory' of what speakers most likely intend by their utterances, rather than a full-scale, elaborated 'prior theory' which would somehow (impossibly) deduce the conditions of possibility for language, interpretation, or communicative utterance in general.

TRAVELLING LIGHT: DAVIDSONIAN SEMANTICS

Hence Davidson's famously cryptic pronouncement (in his essay 'A Nice Derangement of Epitaphs') that 'there is no such thing as a language', or at any rate nothing that could possibly answer to such misplaced philosophical demands.21 Hence also his equally cryptic closing remark in 'The Very Idea of a Conceptual Scheme': that 'in giving up the dualism of scheme and world, we do not give up the world, but reestablish unmediated touch with the familiar objects whose antics make our sentences and opinions true or false'.22 Such

– one could argue – is the end of the road that analytical philosophy has been travelling over the past five decades and more. It is the story, familiar from Quine, which starts out by collapsing the logical-empiricist distinction between analytic and synthetic statements (or 'truths of reason' and 'truths of fact'), and which then goes on to argue – in holist fashion – that 'our statements about the external world face the tribunal of sense experience not individually but only as a corporate body'.[23] This squares well enough with the above-quoted passages from Davidson, inclining as they do toward a pragmatist outlook where 'knowing a language' is the same thing as 'knowing a truth-theory', but where both can be more simply cashed out in terms of 'knowing one's way about the world'. It also matches his suggestion that 'truth' may be construed for all practical purposes – along with 'language' and other such otiose items – simply by adjusting the Tarskian account so as to leave room for 'passing theories' which converge on the attitude of 'holding-true' or the *ad hoc* ascription of speaker's belief.

This is why Rorty, Malpas and others of a kindred persuasion can plausibly lay claim to Davidson as an ally in the cause of 'post-analytic' philosophy. That is, they can cite his texts as further evidence that a formalized (Tarskian) theory makes no substantive contribution to defining the structure or the content of truth, and should therefore be regarded as at best redundant – in the technical sense of that term – or else as a mere *reductio ad absurdum* of the whole analytical project. Davidson himself yields ground to this reading when he adopts a minimalist line in semantics ('there is no such thing as a language') and a maximalist – or holistic – approach to the issue of interpretive context (i.e., that true beliefs are those which hang together with the entire set of such beliefs accepted by some given speaker at some given time). This follows from his urging that we abandon the quest for any theory – any 'prior theory' of truth or radical translation – that would take language as its object, rather than sensibly resting content with the kind of intuitive ad hockery that amounts to the linguistic equivalent of just 'knowing our way about in the world'. His aim in all this is of course to cut out the idea of conceptual schemes (language-games, paradigms, discourses, etc.) since such thinking leads on to those varieties of framework-relativism which Davidson firmly rejects. But it is hard to see how this purpose can be served by an argument that finds no room for language, propositions, sentences or statements as bearers of determinate truth-values, and which has to rely

on such ill-defined notions as that of a 'passing theory', itself conceived none the less as somehow putting us directly in touch with
those 'familiar objects whose antics make our sentences and
opinions true or false'. For this is to take such a primitive (quasi-be-
haviourist) view of meaning and reference that truth might just as
well drop out altogether, along with Davidson's entire set of claims
as regards the priority of truth-values for a theory of radical translation. And from here it is no great distance – as in Quine – to that
doctrine of full-fledged meaning-holism which relativizes 'truth' to
the open-ended range of stimulus–response situations in which a
speaker might be taken as offering evidence for this or that construal of his/her beliefs taken in the large.

Nor is there much help to be had from Davidson's idea of 'passing
theories' as a means of figuring out speaker's intent on the basis of
various contextual cues and clues, along with a generalized 'principle of charity' that counts them (interlocutors or native inform-
ants) most likely 'right on most matters'. For in the absence of at
least *some* 'prior theory' – some grasp of the relation between sense
and reference, meaning and truth, propositions and real-world
(veridical) states of affairs – there is nothing that could count as a
criterion for language or communicative utterance, as distinct from
just random noises produced in response to some equally random
(because unspecifiable) range of physical stimuli. This is what results from Davidson's conjoining a crudely reductive (behaviourist) account of speaker's intent with a Quinean doctrine of
meaning-holism which lacks – and indeed makes a point of renouncing – any appeal to intermediary items of linguistic (logico-
semantic) sense such as sentences, statements, or propositions.

In 'A Nice Derangement of Epitaphs', Davidson applies this argument to the test-case of malapropisms, that is, the class of utterances which, if literally understood, make no kind of sense and which
therefore *de facto* cannot be construed in accordance with any prior
(e.g., logico-semantic or generalized linguistic) theory. And yet we
can and do come up with plausible ways of interpreting such
utterances, as for instance with a stage personage like Sheridan's
Mrs Malaprop or the nonsense-poetry of Edward Lear and Lewis
Carroll. (Davidson also has some nice examples from the American
sitcom character Goodman Ace, to which might be added – for
benefit of British readers – the inspired aphasic ramblings of 'Pro-
fessor' Stanley Unwin.) We do so, he suggests, most often by picking up localized contextual cues or making some *ad hoc* conjecture

based on their observable (no matter how wild or zany) habits of associative linkage. Moreover, it is the same sort of *ad hoc* guess-work (or 'passing theory') that we standardly bring to bear in cases of metaphor, creative neologism, idiolectal variation, peculiarities of local usage, or other such instances of (to us) unfamiliar speech. Likewise, Davidson argues, when we encounter new additions to our stock of proper names or referring expressions. For again these are items that find no place in our pre-existing range of interpretive competence or lexico-semantic grasp. Thus we have no choice but to interpret them – on first acquaintance – as instances of pure nonce-usage, *hapax legomena* whose meaning can only be divined by some happy combination of intuitive guesswork, hunches about context, practical savvy, and generally 'knowing one's way about in the world'.

This illustrates very clearly the confusions that Davidson gets himself into by retreating from a strong truth-conditional theory of meaning to a minimalist semantics where truth drops out except as defined in holistic or contextualist terms. For on the view put forward in 'A Nice Derangement' it *simply doesn't matter* whether an expression makes sense – or can be assigned some sensible interpretation – in keeping with the ground-rules of linguistic com-petence, the semantic structure of a language, or any other such component of a 'prior theory' taken as exerting some constraint upon the range of well-formed (meaningful) utterances. Moreover, there seems no reason to suppose – on Davidson's minimalist-semantic account – that language (or the speaker's and the listener's shared possession of a language) need play any role what-soever in the process of securing communicative uptake. For if 'passing theories' are in fact all we need – and if these can always override 'prior theories' to the point of guessing what a speaker has in mind on the evidence of his or her gestures, mannerisms, beha-vioural dispositions, verbal tics, responses to stimuli, and so on – then it follows that 'language' is indeed an otiose concept, and that interpretation can get along just as well in the absence of articulate meaning. Of course, as Davidson knows, this flies in the face of much that is taken for granted not only by philosophers and lin-guists but also by speakers when they reflect (in a commonsense or intuitive way) on what is involved in the act of communicative utterance. Thus on the usual view, as he describes it, 'in the case of language the hearer shares a complex system or theory with the speaker, a system which makes possible the articulation of logical relations between utterances, and explains the ability to interpret novel utterances in an organized way'. But he thinks this view altogether mistaken since it involves the appeal to abstract or notional entities ('languages', 'systems', 'logical relations', 'conceptual schemes' and so forth) which serve no genuine explanatory purpose and should hence be consigned – in good pragmatist fashion – to the junkyard of obsolete ideas.

In short, '[w]hat two people need, if they are going to understand one another through speech, is the ability to converge on passing theories from utterance to utterance'. And again: 'we can give content to the idea of two people "having the same language" by saying that they tend to converge on passing theories; degree or relative frequency of convergence would then be a measure of similarity of language'.[24] Anything more – like the demand for some theory of language that could give substantive content to the notions of meaning, reference, and truth – is now viewed by Davidson (like Wittgenstein before him) as just another symptom of that misplaced philosophical 'craving for generality' which creates problems where no real problems exist. For there are, he surmises, 'no rules for arriving at passing theories, no rules in any strict sense, as opposed to rough maxims' ('Nice Derangement', p. 173). In which case we should be willing not only to abandon what Davidson calls 'the ordinary notion of a language', but also to erase the distinction between 'knowing a language' and 'knowing our way around in the world generally'. For it is just that distinction – on his account – which has produced the idea of language (or of sentences, statements, propositions, etc.) as standing in some more-or-less accurate relation of correspondence with factual states of affairs. And this idea has in turn given rise to all the well-known problems about explaining how such 'facts' could ever be construed except under some linguistic description, or again, how language could be prised off the world so as to decide whether statements match up with the way things stand 'in reality'. And so the way lies open to framework-relativism, to talk of incommensurable paradigms, language-games, discourses, 'conceptual schemes', and sundry variations on the same line of argument.

However, Davidson's solution seems a desperate remedy if it entails giving up any notion of 'knowing a language' except in so far as such knowledge comes about through the *ad hoc* convergence of speaker's and interpreter's meaning from one utterance to the next. 'A passing theory really is like a theory in this at least, that it is derived by wit, luck, and wisdom from a private vocabulary and grammar, knowledge of the ways people get their point across, and rules of thumb for figuring out what deviations from the dictionary are most likely' ('Nice Derangement', pp. 173–4). Still one has to ask what such a 'theory' could amount to, given the highly restrictive conditions placed upon it by Davidson's minimalist semantics and his requirement that the idea of 'knowing a language' be reduced accordingly to 'wit, luck, and wisdom' plus whatever the speaker happens to possess of 'private' (idiolectical) resources. For if one takes this description at anything like face-value then it leaves no work for the 'theory' to do, aside from contributing – in some vague and unspecified way – to a process of intuitive guesswork which could just as well function in the absence of language, that is to say, through observation of the various empirical cues and clues provided by a speaker's overt behaviour in this or that context of utterance. If we wish to give content to the notion of two people 'having the same language' then we can do so least problematically, Davidson suggests, 'by saying that they tend to converge on passing theories'; in which case 'degree or relative frequency of convergence would be a measure of similarity of language' (p. 173). But he thinks of this as just a minor concession to received (philosophical or 'commonsense') ideas of what goes on in the act of linguistic communication. For really there is as little use for notions like 'the same language' (or generalized accounts of shared speaker-competence) as there is for explanations of communicative uptake that rest upon prior – as distinct from purely *ad hoc*, transient or 'passing' – theories.

Thus on Davidson's view, quite simply, 'the asymptote of agreement and understanding is where passing theories coincide' (p. 169). But such convergence cannot itself become the basis for a generalized theory of linguistic competence, semantic structure, utterer's meaning, or whatever. No doubt there is a sense – a trivial sense – in which two speakers who 'have the same language' must to that extent share some prior theory (or some overlapping range of such theories) with which to make a start in the process of mutual comprehension. Nevertheless it is the case, Davidson argues, that in order to interpret novel or other than routine (protocol) sentences they will need to make use of a passing theory – a mixture of 'wit, luck, and wisdom' – which pertains to that singular context of occurrence and which can therefore neither be *derived from* nor *applied to* any larger (transcontextual) theory of communicative utterance in general. 'Not only does it [the passing theory] have its changing list of proper names and gerrymandered vocabulary, but it includes every successful – i.e., correctly interpreted – use of any other word or phrase, no matter how far out of the ordinary' (p. 169). This makes it sound most like the procedure with word-processor spellcheck programs where the software signals some anomaly in the text, some lexical item that cannot be discovered in its dictionary. One then has to choose between accepting the proposed alternative – most often a wildly unsuitable word selected on grounds of alphabetical adjacency or vague semantic collocation – or, if one prefers, ignoring the suggestion and adding the new term to its customized memory.

At any rate this seems fairly close to what Davidson has in mind when he argues that 'correct' interpretation has much more to do with 'passing' than with 'prior' theories. For on his view the test of any such theory is its capacity to cope with an open-ended range of novel expressions and contexts, a capacity which finds no analogue in the various abstract or idealized models of linguistic competence. On the standard view, conversely,

> what is essential to the mastery of a language is not knowledge of any particular vocabulary, or even detailed grammar, much less knowledge of what any speaker is apt to succeed in making his words and sentences mean. What is essential is a basic framework of categories and rules, a sense of the way English (or any) grammars may be constructed, plus a skeleton list of interpreted words for fitting into the basic framework. ('Nice Derangement', p. 171)

But, according to Davidson, this offers no help in explaining how it is that we can and do manage to negotiate all manner of hitherto unmet-with speech situations, contexts of utterance for which there exists no adequate generalized theory. Such projects fail 'for the same reasons the more complete and specific prior theories fail: none of them satisfies the demand for a description of an ability that speaker and interpreter share and that is adequate to interpretation' (p. 171). For in order to be 'adequate' in Davidson's (minimalist-semantic) sense such a theory must in fact be as little like a 'theory' as possible, at least on the normal definition of that term assumed by linguists and philosophers.

Hence his counter-argument to the above set of claims, taken as typifying most of what passes for 'adequate' theory (or the aspiration thereto) in present-day linguistics and philosophy of language. 'Any general framework', Davidson writes,

> whether conceived as a grammar for English, or a rule for accepting grammars, or a basic grammar plus rules for modifying and extending it – any such general framework, by virtue of the features that make it general, will by itself be insufficient for interpreting particular utterances. The general framework or theory, whatever it is, may be a key ingredient in what is needed for interpretation, but it can't be all that is needed since it fails to provide the interpretation of particular words and sentences as uttered by a particular speaker. In this respect it is like a prior theory, only worse because it is less complete. (p. 171)

This appears to make at least one significant concession to the standard view. For there may exist some 'general framework' – some prior set of rules, grammatical constraints, derivative forms, structures of semantic representation – that would serve (so to speak) as a normative background against which to register any novel expressions encountered in the course of everyday talk. However, by the end of this passage, as so often in his essay, Davidson has deftly taken back with one hand what he seemed to hold out in the other. That is to say, he has reverted to the minimalist line which maintains that prior theories are always pretty useless – and 'general frameworks' even more so – when it comes to giving an adequate account of what transpires in actual cases of communicative uptake. Thus the notion of a framework is 'worse' ('less complete') than that of a prior theory since in purporting to explain a lot more about language – or to do so at a higher theoretical level – it explains next to nothing about the way that particular speakers get their meaning across in particular contexts of utterance.

What is so remarkable about Davidson's essay is its way of coolly, commonsensically advancing proposals which go clean against both the ordinary (commonsense-intuitive) understanding of language and just about everything that linguists and philosophers have lately had to say in this regard. On at least six basic points it ignores every standard working premise as to what is meant by having (or sharing) a language and what should count as an adequate theory thereof. (That Davidson rejects this whole line of thought as fundamentally misguided since on his account 'there is no such thing as a language' and, *a fortiori*, no such 'adequate' theory to be had is scarcely an argument for counting his proposals immune from criticism on these or other grounds.) Thus: (1) there is good reason, as well as Wittgensteinian warrant, to question his idea of passing theories as constructed by 'wit, luck, and wisdom' out of a *'private* vocabulary and grammar'. For this would render their meaning utterly opaque not only from the interpreter's stand-point but also for the speaker in so far as he or she could then possess no criteria of meaningful (communicative) utterance as distinct from mere subjective whimsy or 'private' association. (2) It is wrong – a fairly basic misunderstanding – to equate the idea of 'knowing [or learning] a language' with some process of *ad hoc*, piecemeal addition such as occurs (*vide* my analogy above) in the two-way (quasi-) interactive exchange between computer program and individual user. And this because (3), as Davidson acknow-ledges in passing elsewhere, language is characterized by recursive structures at the level of syntax and semantics. Moreover (4), it is just this feature of language – as theorized most explicitly by Chomsky and proponents of a transformational-generative ap-proach – that allows for the production and interpretation of a vast (potentially infinite) range of utterances without need for David-son's problematic recourse to a piecemeal additive account.

There is also (5) some evidence of a type/token confusion in his idea that the 'same' word or expression may always require inter-preting anew – by means of an *ad hoc* passing theory – from one context of utterance to the next. It is worth noting here that a similar criticism has been brought against Derrida's reading of speech-act philosophy, understood by some commentators as a version of minimalist semantics much akin to that of Davidson.[25] Thus on Derrida's account the 'iterable' character of performatives – their capacity to function across an indeterminate range of speech-situations – is taken to debar any possible normative appeal to speaker's intentions on the one hand or validating context on the other.[26] This criticism misses the mark with Derrida, as I have argued at length elsewhere.[27] But it does go some way toward explaining why Davidson adopts such a minimalist approach, that is, his idea that 'sharing a language' comes down to just a matter of 'wit, luck, and wisdom', of occasional conver-gence on passing theories, or of interpreting 'particular words and sentences as uttered by a particular speaker'. Hence (6) his curi-ous (and surely mistaken) idea that malapropisms, nonce-usages, verbal parapraxes and suchlike anomalous items may be taken as providing a test-case for theories of communicative utterance in general, or a fair working sample of what normally goes on in the course of figuring out speaker's intent. For this argument supposes a degree of radical meaning-variance from context to context that would require the interpreter to start out almost from scratch in making sense of every utterance that came her or his way.

'Almost', be it noted, since Davidson does allow prior theories to play some part in this process, albeit a walk-on (more aptly: a walk-off) role whose function appears mainly to be that of displaying its own inadequate resources. Such would be the case in those kinds of social context – like our first conversation with some new acquaintance – where as yet we have no passing theories to go on. For '[t]he less we know about the speaker, assuming we know he belongs to our language community, the more nearly our prior theory will simply be the theory we expect someone who hears our unguarded speech to use' (p. 170). But, on Davidson's account, this will barely permit us to make an informed guess as to what that speaker may have in mind when uttering some particular word or sentence outside the range of our own more habitual or routine habits of speech. It is at this point that the passing theory comes in, that is to say, at the stage where our *linguistic* resources (as Davidson defines them) have nothing very useful to offer, and where 'theory' gets along on a minimalist basis of contextual observation plus guesswork. In which case, as Davidson half-rhetorically asks, '[w]hy should a passing theory be called a theory at all?' (p. 170). Why indeed? one might respond since the work it has to do seems hardly to merit that title. What it comes to, in effect, is a pragmatist variant of the Tarskian formalized semantic schema, a rule-of-thumb version on which 'truth' drops out – together with the whole (now redundant) apparatus of theory – just so long as one can keep the conversation going with some such vague assurance in the background. Thus 'the sort of [passing] theory we have in mind is, in its formal structure, suited to be the theory for an entire language, even though its expected field of application is vanishingly small' (p. 170).

Here again we can see just how far Davidson is prepared to go with his occasionalist approach to issues of linguistic understanding. This emerges not only in his strictures on anything that resembles a 'general' theory of language and interpretation, but also in his limiting judgement on 'prior theories' and – beyond that – in his treatment of (so-called) 'passing theories' as really little more than an override-mechanism for clearing those other impediments out of the way. One useful pointer in this direction, Davidson suggests, 'is to reflect on the fact that an interpreter must be expected to have quite different prior theories for different speakers – not as different, usually, as his passing theories; but these are matters that depend on how well the interpreter knows his speaker' (p. 171). From which it follows that there must be as many 'languages' – or as many ways of 'knowing a language' even within what nominally counts as the 'same' linguistic community – as there are individual speakers, interpreters, and localized contexts of utterance. This is why Davidson can plausibly (by his own lights) reduce the whole business of 'correct' interpretation to a matter of 'wit, luck, and wisdom', or of one's happening to hit upon the right meaning – that is, what the speaker truly had in mind – quite aside from what his or her utterance *meant* as construed according to some otiose theory of semantic structure, propositional content, grammatical competence, or whatever. It also helps to explain – if not justify – his narrowing the scope of any useful interpretation-theory to the point where it would treat every sample utterance as (for all that we can know beforehand) a one-off, wholly context-specific, or uniquely-occurring expression. The upshot of this, on Davidson's view, is that we are always to a large extent working from scratch like a first-time audience exposed to the effusions of Sheridan's Mrs Malaprop, or again, like a literary critic confronted with some passage from 'Jabberwocky' or *Finnegans Wake*.

Thus what Davidson calls 'interpretation' should perhaps better be construed as a form of applied cryptoanalysis, an exercise in discovering how far we might get if deprived of all the standard interpretive resources that go toward 'knowing one's way about' in language. This approach – if taken on its own express terms – would discount just about everything pertaining to the interpreter's prior knowledge in order to provide for those special cases (or, as Davidson would have it, those not-so-special cases) that arise through a speaker's ignorance, error, or failure to straightforwardly say what they mean. 'Consider the malaprop from ignorance', Davidson invites us, as proof of this contention that 'most of the time prior theories will not be shared', and moreover that 'there is no reason why they should be'.

Mrs Malaprop's theory, prior and passing, is that 'A nice derangement of epitaphs' means a nice arrangement of epithets. An interpreter who, as we say, knows English, but who does not know the verbal habits of Mrs Malaprop, has a prior theory according to which 'A nice derangement of epitaphs' means a nice derangement of epitaphs; but his passing theory agrees with that of Mrs Malaprop if he understands her words. (p. 170)

His argument here raises so many problems that it is hard to know where to begin. But it is clear – so far as anything is clear in all this – that what Davidson means by 'understanding her words' has little or nothing to do with those words (or any proper 'understanding' of them). Rather, it is a question of playing it by ear, ignoring the sense (whether manifest or latent) of pretty well everything she says, and instead trusting to chance associations at the phonetic and – less often – the semantic level. (Again there is a close analogy here with the frequently quite random and absurd suggestions thrown up by a computer spell-check program.) This is why, at the limit, Davidson's theory of interpretation can bypass language altogether and appeal directly to speaker's meaning – or intention – as construed on an all-purpose 'principle of charity' designed only to optimize the sense in this or that context of utterance. But then what is the point of invoking an interpreter 'who, as we say, knows English', or of continuing to talk about 'theories' (prior or passing) which can have no real utility or purpose as applied to the case in hand? All that such talk amounts to, in the end, is a holding-operation which prevents 'theory' (and 'language' along with it) from dropping completely out of view, while clearing the way for a minimalist semantics – or a thoroughgoing pragmatist approach – that would render those notions altogether redundant.

SOME PROBLEMS WITH ANTI-REALISM

I have rather belaboured Davidson's argument since it strikes me as symptomatic of much that is wrong with what currently passes as 'post-analytical' philosophy. That is to say, it takes for granted the impossibility of justifying truth-talk in other than formal (Tarskian) terms; the lack of any viable alternative account based on substantive (*de re*) as opposed to purely semantic (*de dicto*) conceptions of meaning and truth; and hence the necessity – as Davidson sees it – for a doctrine of full-scale meaning-holism which offers no guidance in construing particular speech-acts, and which effectively treats them as so many items of sheer, unexampled nonce-usage. Such is, presumably, the point of his remark that the kind of theory in question 'is, in its formal structure, suited to be the theory for an entire language, even though its expected field of application is vanishingly small' ('Nice Derangement', p. 170). For on Davidson's view it must always be subject to a law of rapidly diminishing returns, that is, to the inverse relation that holds between a theory's purported explanatory scope and its usefulness as a practical 'rule of thumb' with application to particular contexts of utterance. Whence the idea – uncongenial to Davidson but often derived from his writings – that we have entered not only a 'post-analytic' but also, where longer perspectives are invoked, a 'post-philosophical' culture. (This idea has understandably gained more ground among literary and cultural theorists than among professional philosophers.) What emerges most clearly in Davidson's case – from 'The Very Idea of a Conceptual Scheme' to 'A Nice Derangement of Epitaphs' – is a version of the background narrative implicit in many such claims to the general effect that analytic philosophy is a failed enterprise, that it has produced only dead-end or manifestly trivial results, and should therefore be abandoned in favour of other, more productive or 'edifying' projects.

This story has achieved its widest circulation through Richard Rorty's deft and vivid rendition in his book *Philosophy and the Mirror of Nature*.[28] Here it takes the form of a narrative survey covering most of the major episodes in Western philosophical tradition from Plato and Aristotle to Descartes and Kant and thence – *via* Hegel – to (among others) Husserl, Heidegger and Gadamer on the modern 'continental' side and Frege, Russell, Quine, Davidson, Putnam, and Kripke as avatars of the broadly 'analytic' line. The narrative is angled at every stage toward Rorty's wished-for *dénouement*. It brings together a deflationary account of 'truth' in the Tarski-Davidson mode with a pragmatist covering approach, that is, the idea that truth *just is* what currently and contingently counts as 'good in the way of belief'. Quine is enlisted as a handy authority for the case against any lingering attachment to 'truths of fact' or 'truths of reason', that is to say, those tenacious 'dogmas of empiricism' grounded in the old (whether Kantian or logical-empiricist) distinction between synthetic and analytic statements.

Those who betray such residual inclinations – among them causal realists like Kripke and half-way converts like Putnam – are treated with a kind of pitying fondness as having not yet realised (even after Quine and Wittgenstein) that these are just language-games or styles of talk whose sell-by date has long since expired.[29]

Hence the great merit of Davidson's work, as Rorty sees it: to have pushed analytic philosophy to the point where 'truth' becomes either a product of purely formal (circular) self-definition, or – what amounts to much the same thing – a choice among the various language-games, discourses, or vocabularies currently on offer. I have argued elsewhere that Rorty gets Davidson wrong as regards his earlier position in 'The Very Idea of a Conceptual Scheme'.[30] Here one finds at least the outline of a more substantive argument with respect to the role of a truth-theory as applied to issues of utterer's meaning and communicative grasp. But in 'A Nice Derangement of Epitaphs' Davidson presents a much weakened version of this argument which shows why Rorty – along with other commentators like Okrent and Malpas – can promote the otherwise improbable alliance between Davidsonian ('post-analytic') and Heideggerian (depth-ontological) kinds of talk. For if the analytic concept of truth turns out to be trivial or redundant – as it does, at least arguably, on Davidson's later account – then the way is wide open for these writers to suggest that we require a wholly different, more 'primordial' conception which has somehow been occluded by the entire history of 'Western metaphysics' from Plato and Aristotle to Descartes, Leibniz, Kant and their latter-day progeny. This approach would view analytical philosophy as merely the latest, most extreme or symptomatic showing of that deep-laid oblivion – or forgetfulness of Being – which has characterized the past two millennia of Western thought. It would thus be receptive to Heidegger's talk of truth as *aletheia* (unconcealment), as a renewed openness to Being, or as a questing-back for those profound intimations – to be glimpsed obscurely in the extant fragments of the pre-Socratics – whose meaning eludes any possible treatment in formal or logico-semantic terms.[31]

So it is that Rorty can present his whole narrative of Western philosophy to date as a history of errors – of repeated wrong turns – brought about by the otiose (epistemological) idea of mind as a 'mirror of nature', a 'glassy essence' which should properly reflect how things stand in the order of reality or the order of clear and perspicuous mental representations. For these are just metaphors taken as concepts, 'truths' (according to the much-quoted passage from Nietzsche) whose figural origin has now been forgotten and which can thus delusively pass themselves off as literal, transparent, or veridical ideas.[32] From all of which it follows – on this Rortian account – that we can best throw off the dead weight of inherited philosophic concepts by taking Davidson (along with Quine and others) to have shown up their chronic obsolescence, and by looking to Nietzsche, Heidegger, and Derrida (not to mention 'private ironists' like Foucault and 'strong revisionists' like the literary critic Harold Bloom) for some idea of what philosophy might yet become once shorn of its delusive epistemological pretensions. For it could then assume its rightful – though strictly non-privileged – role as one more voice in the 'cultural conversation of mankind', a voice that lays claim to no special expertise in matters of truth, reason or right understanding. Its various contributions would henceforth be valued in so far as they offered new kinds of metaphor to live by, better (more persuasive) depth-interpretations of our communal being-in-the-world, or again – as with Foucault and Bloom – novel possibilities for private-creative self-description. Such (Rorty thinks) is the best – indeed the only – prospect for rescuing philosophy from its present condition as a marginalized discipline endlessly embroiled in pseudo-problems of its own artificial creation.

Davidson is clearly not altogether happy with this reading of his work as yet another nail in the coffin of old-style 'constructive' philosophical debate. At times he goes *almost* so far as to declare that Rorty has quite simply got him wrong, and that issues of interpretive meaning and truth cannot be reduced *tout court* to what counts as 'good in the way of belief'.[33] However, one can see from his various (often bewildering) changes of tack in response that Davidson is hard put to provide any strong counter-argument – or corrective self-interpretation – by way of adequate rejoinder. Thus on other occasions he appears largely willing to view his own work through the proffered Rortian lens and accept that Tarski's schema is indeed nothing more than a roundabout route, *via* minimalist semantics, to the looked-for pragmatist outcome.[34] Still there is no reason, foregone preference aside, to endorse either Rorty's general diagnosis or those aspects of Davidson's thinking which lay themselves open to deflationary treatment in this manner. That is to say, it is only on a certain (very partial) account of analytic philosophy and its discontents that this narrative turns out to support the idea of a terminal crisis whose origins lie far back in the history of Western thought, and whose upshot is the choice between an empty formalism and a 'deep' (hermeneutical or 'edifying') quest for alternative sources of wisdom.

Davidson himself seems intermittently aware that this is a false dilemma, one that has come about chiefly as a consequence of those well-known problems encountered with the logical-empiricist programme and its various successor doctrines. Thus on the Rortian (or 'post-analytic') view there is simply no hope of reviving that old enterprise once we have taken the point of (for instance) Quine's 'Two Dogmas of Empiricism', Nelson Goodman's restatement of the Humean puzzles about induction, and the many criticisms that have lately been brought against the covering-law (deductive-nomological) approach to issues in epistemology and philosophy of science.[35] Again, Davidson is less inclined than Rorty to treat all this as a ready-made pretext for junking the 'mainstream' (analytic) tradition and launching out into new, more adventurous seas of thought. Hence – as I have noted – his scattered remarks about the inadequacy of a purely Tarskian (disquotational) theory of truth and the consequent need to fill out that theory with something more substantive or at any rate not just trivially self-confirming. But in the end Davidson always comes back to some version of the standard Quinean trade-off between, on the one hand, a pragmatist or holist conception of 'truth' as the sum-total of existing beliefs and, on the other, a minimalist semantics where 'passing theories' – or inspired guesswork – are the best we have to go on in matters of localized or detailed interpretive grasp.

This is why Rorty, Malpas and others can so easily present him as a half-way convert to their own, jointly pragmatist and depth-hermeneutic approach. No matter that Davidson's preferred idiom – his typically casual, laid-back talk of passing theories, rules of thumb, 'wit, luck, and wisdom' and the like – is about as remote as can be from Heidegger's portentous style of philosophizing, early and late. Rorty gets over this problem to his own satisfaction by recommending that we simply ignore all that stuff about *Dasein*, Being, 'Western metaphysics', and so on, or treat it with a pragmatist pinch of salt as just another language-game [or 'final vocabulary'] which may have some short-term edifying uses quite aside from its proclaimed depth-ontological import. My point is that Davidson invites such readings – or puts up minimal resistance to them – in so far as he adopts a holist view which leaves no room for any theory of truth, meaning, or interpretation except the sort of wholly uninformative theory which *either* goes redundant (in the Tarskian manner) or cashes out as a pragmatist idea of what's 'good in the way of belief'. And in this case, so Malpas argues, there must be *something more* to Davidson's theory, some further (implicit or by him unrecognized) dimension which can be revealed only through a mode of Heideggerian depth-ontological enquiry. For otherwise that theory is so limited in scope – so devoid of substantive or explanatory content – that the exegete is hard put to explain why it has attracted such widespread interest.

Malpas puts the case as follows in a passage that shows how readily – or with how few minor adjustments of emphasis and idiom – Davidson can be nudged in a pragmatist and, beyond that, a Heideggerian direction.

> While the idea of truth as *aletheia* ['unconcealment'] might at first sight seem alien to the Davidsonian account, it is, in fact, suggested in the very methodology which Davidson employs. It is a methodology in which understanding develops through our dialogue and involvement with other speakers and entities within the world. *Aletheia* is the event of opening up – of freeing up – which makes such dialogic intervention, within which understanding arises, possible. . . . Dialogue is thus itself constituted as 'truthful', not only in that it can give rise to truth as correspondence or as coherence, but in that dialogue is itself constituted as a constant process of disruption – in which new possibilities of meaning are revealed – and reconsolidation in which certain of those possibilities are taken up and, as they are taken up, others are concealed. Thus dialogue reveals and conceals; truth is the revealing-concealing itself.[36]

One cannot imagine Davidson responding with much enthusiasm if confronted with an exegesis of his work in this *echt*-Heideggerian mode. At the very least he might seek answers to the following fairly obvious questions. First: how can truth-as-correspondence (*adaequatio verbi ad rem*) be taken as in any way arising from a dialogue where 'entities' emerge only by virtue of their playing some significant role in that same dialogue? The Heideggerian doctrine may perhaps seem plausible if linked with a coherence-theory of truth but it reduces to manifest nonsense when applied to (or tested against) any version of the correspondence-theory.

Second: what sense can be made of this idea that such 'entities' depend for their very existence on a process of obscurely conjoint 'revealing-concealing' whose history – as Heidegger construes it – is coterminous with that of 'Western metaphysics' and its epochal evasions of the question of Being? Whatever the difficulties with Davidson's account they are scarcely resolved – or rendered more perspicuous – by translation into a Heideggerian mode of depth-ontological talk. And third: how can a theory of truth-as- coherence – which is really what all this amounts to – make allowance for the sorts of disruptive 'intervention' which Malpas here envisages? For this can occur only in consequence of our sometimes asking new questions and gaining new knowledge of particular, well-defined 'entities', and not through such vastly generalized (and arguably vacuous) questions as Heidegger's 'What Is a Thing?'.[37]

In science it is often a matter of discovering some deep (micro-structural) property of the item in question which enables us better to describe and explain its various physical manifestations. This applies most strikingly in the field of modern particle-physics where for instance, as Newton-Smith remarks, the term 'electron' served as 'a predicate . . . introduced with the intention of picking out a kind of constituent of matter, namely that responsible for the cathode-ray phenomenon'.[38] One could multiply examples to similar effect across the range of scientific research, whether at the micro- or the macro-physical level. In the former case science has typically progressed from a stage of rational conjecture – or speculative thought-experiment – to a stage at which it is able to verify the existence of ever more elusive or recondite subatomic particles. That is to say, they start out as theoretical constructs whose reality can then be confirmed (or disconfirmed) through more refined observational techniques or more advanced experimental procedures.[39] In the latter case also – in the macro-domain, as for instance with astrophysics – knowledge accrues by seeking out phenomena which permit the testing of some powerful but hitherto unproven hypothesis in terms of its causal-explanatory yield. Thus the theory of an expanding universe received strong support from observation of the red-shift – the statistically abnormal predominance of certain wavelengths in the colour-spectrum – whose presence could best be accounted for on the premise that they issued from receding galaxies.

These advances are achieved very often with the advent of some new technology, such as – to take the most relevant examples –

radar telescopes or the Wilson cloud-chamber which first made visible the tracks of subatomic particles passing through it. Recent examples would include the development of more powerful particle-accelerators and electron-microsopes with higher powers of resolution.[40] But they can also come about – as Karl Popper has emphasized in his book *Quantum Theory and the Schism in Physics* – through thought-experiments (such as those conducted by Einstein and Bohr in their famous series of debates) which cannot be carried out physically for whatever reason but which none the less offer strong support for some hypothesis concerning real-world entities, processes, or events.[41] Of course such results remain open to challenge, as Popper is naturally quick to concede given his well-known fallibilist view that the mark of genuine scientific truth-claims – as opposed to pseudo-scientific dogmas – is their capacity for falsification when exposed to rigorous critical testing.[42] Thus for instance he attaches considerable weight to a number of such thought-experiments – among them the classic Einstein–Podolsky–Rosen conjecture – which Popper interprets as supporting a realist (as opposed to a radically indeterminist) construal of the theory of quantum-mechanics. On the received view, as Popper acknowledges, Bohr got the better of these arguments with Einstein and established a number of theses – comprising the so-called 'Copenhagen interpretation' – which he (Popper, like Einstein) finds altogether unacceptable.[43] Nevertheless there is a sense in which *any* conjecture arrived at on the basis of such a thought-experiment must claim to have discovered some truth (however puzzling or counter-intuitive) about the way things stand in reality. And this applies even to those cases – like the celebrated 'two-slit' experiment and its various latter-day refinements – which are often taken to count against a realist construal of quantum mechanics. For in so far as they possess any genuine demonstrative or probative force – any bearing on the question what would *actually occur* if that set-up were realised in physical form – they are *de facto* committed to some version of the correspondence-theory which holds statements true (or hypotheses justified) in virtue of their real-world descriptive and explanatory merits.[44]

Thus, according to Popper, it cannot be an argument against realism in the quantum domain that these debates took place at a 'purely speculative' level where observational proofs were not (or not yet) to be had and where rival hypotheses could be tested only through the conduct of thought-experiments. For it is still the case 'that these metaphysical speculations *proved susceptible to criticism* – that they could be critically discussed'. Moreover, this was a discussion 'inspired by the wish to understand the world, and by the hope, the conviction, that the human mind can at least make an attempt to understand it'.[45] Thus Popper can himself criticize certain aspects of the thinking of Bohr, Heisenberg and others – doctrines which, if followed consistently, would preclude any further scientific advance – while professing the highest regard for their actual achievements in the quantum-theoretical field. On his view these doctrines were also 'metaphysical' but in a different (pejorative) sense of that term. That is to say, they raised currently-existing problems to a high point of obscuranist dogma, a refusal to acknowledge that those problems might yet be resolved by further thought-experiments or perhaps by the discovery of so-called 'hidden variables' at a deeper, more complete or adequate level of explanation.[46]

Indeed, it was Bohr's contention – supported by Heisenberg – that the quantum-theory was already 'complete' in all its most important details, and hence (on *a priori* grounds) that no such solutions could ever be forthcoming. Science had now arrived at the absolute limit of its capacity to describe, explain, or conceptualize phenomena in the microphysical domain. For those phenomena must still undergo translation into the language of 'classical' (pre-quantum) physics, this being – as Bohr understood it – the only possible language or descriptive framework wherein to represent them in scientifically (or humanly) intelligible terms.[47] Thus:

> when he [Bohr] accepted quantum mechanics as the end of the road, it was partly in despair; only classical physics was understandable, was a description of reality. Quantum mechanics was not a description of reality. Such a description was impossible to achieve in the atomic region; apparently because no such reality existed: the understandable reality ended where classical physics ended. The nearest to an understanding of atoms was his own principle of complementarity.[48]

Thus Bohr came up with a set of sharply paradoxical theses concerning the gulf – the ontological-epistemic gap – that would *always necessarily* exist between quantum 'reality' and what could be observed, known, or stated about it. He concluded that nothing could resolve the dualism of wave and particle theories; that it was (and must remain) strictly impossible to conduct simultaneous accurate measurements of a particle's position and velocity; that any attempted measurement of either value would create interference-effects (as for instance through the impingement of photons upon particles whenever such an act of observation took place); and that the so-called 'reduction of the wave-packet' – that is, its 'jumping' into discrete wave or particle form – must itself be regarded as an absolute limiting condition, an instance of the quantum uncertainty-principle that always obtained in the realm of subatomic physics.[49]

Hence what Popper justifiably describes as the outlook of 'despair' to which Bohr was driven by his insistence that the quantum-theory was 'complete' as it currently stood. For this thesis had the twofold ironic consequence: first, that such a theory was *incomplete* by very definition (since it posited a realm of noumenal entities forever beyond reach of observation, measurement, or determinate knowledge); and second, that science lacked even the means of adequately describing this predicament since it disposed of no language – no conceptual framework – other than that of a (perhaps slightly modified) 'classical physics'. On the one hand any notion of quantum 'reality' dropped out of the picture, giving way to a radically subjectivist (observer-relative) interpretation. On the other, there developed an extreme version of the linguistic incommensurability-doctrine which maintained that quantum-theoretical ideas *could only* be explained in the classical language, but would then of necessity be subject to such errors and distortions as to render that exercise well-nigh meaningless. And so it came about, according to Popper, that a 'whole generation' of younger physicists followed Bohr and Heisenberg in adopting the patently 'absurd view' that 'one can learn from quantum mechanics that "objective reality" has evaporated'.[50]

The same lesson has very often been drawn by thinkers in other fields – from philosophy of science to ethno-linguistics and cultural or literary theory – keen to enlist scientific support for their relativist and anti-realist views.[51] What unites them is the shared belief that 'reality' and 'truth' are values contingent upon our adopting some particular descriptive framework or conceptual scheme, and hence that any shift in the order of discourse will bring about a wholesale paradigm-change which leaves us quite literally living 'in a different world'. Such is Thomas Kuhn's well-known account of what happens during periods of 'revolutionary' science when all the old certitudes are swept away and there exists no reliable means of translation between theories, postulates, observation-statements, or even – at the limit – between *objects* as construed in pre- and post-revolutionary discourse.[52]

I have already discussed Davidson's argument, in 'The Very Idea of a Conceptual Scheme', to the effect that these and similar claims are manifestly self-refuting. That is, they fall prey to performative self-contradiction in so far as they purport to establish the existence of incommensurable paradigms, discourses, or conceptual schemes on the basis of comparisons and contrasts which *could not be made* if the thesis were accepted in anything like its strong (e.g. Quinean, Kuhnian, or Whorfian) form. There is a kindred, albeit more complex and oblique line of argument in Derrida's essay 'The Supplement of Copula'.[53] Here he disputes certain claims advanced by the linguist Emile Benveniste with regard to the priority of language over thought, or – in so far as the issue can be framed in disciplinary terms – of 'linguistics' over 'philosophy'. Thus for Benveniste it is a matter of demonstrating that even the most (to us) 'elementary' structures of logical thought are in fact derivative from a language – the ancient Greek – whose lexico-grammatical resources have exerted a deep and continuing hold upon the entire tradition of Western philosophic thought.[54] From which he concludes that this particular range of naturalized (language-based) categories, logical distinctions, forms of predicative judgement and so forth have simply been mistaken, by philosophers from Aristotle down, as 'laws of thought' that necessarily transcend all mere particularities of cultural or geo-historical place and time. What Derrida remarks, by way of counter-argument, is Benveniste's reliance on those same (strictly indispensable) concepts and categories which provide the only possible means of construing his various items of linguistic evidence and thus making sense of his own thesis with regard to their divergent ontological schemes. Like Davidson, he (Derrida) finds it strictly impossible to maintain such a thesis in anything like its full-fledged relativist form. And the same applies to Benveniste *despite and against* his express avowal of precisely that doctrine, that is, his claim that there exist other languages that possess no equivalent to the predicative function or the logico-grammatical copula 'A is B'.

It is well to be clear as to the pertinence and force that such arguments properly possess. Derrida is *not* for one moment suggesting that Benveniste's work stands or falls on the strength of a thesis

– the language-dependent status of reason, logic, categorical judgements, grammatico-predicative structures, and so forth – which
proves to be demonstrably self-refuting, and which thus invali-
dates the entire enterprise. Rather, he seeks to distinguish this
strain of somewhat facile ultra-relativist argument from the
genuine insights that Benveniste achieves despite – and against –
his espousal of a doctrine that would seemingly render such insights impossible. As I have said, Derrida's line of reasoning to
some extent falls square with Davidson's, that is, in so far as it
appeals to certain strictly *indispensable* criteria (of logic, consistency,
and self-reflexive application) which are presupposed in any theory of language and also, by extension, any first-order natural
language interpretable on such a theory. But Derrida's case is more
convincing than Davidson's – more resistant to Rorty-style pragmatist cooption – by virtue of its not yielding ground to a minimalist semantics (or a holistic theory of meaning and truth) which can
then be coaxed down to the level of straightforward consensus-belief. For it is Davidson's half-way, on-and-off willingness to
concede this possibility which leaves him with little defence against
readings of his work that would count all the truth-talk pretty
much redundant, or at most just a brief (self-cancelling) detour on
the route to that pragmatist conclusion.

I have argued that such readings are less plausible – or at any rate
encounter more resistance – when applied to an essay like 'On the
Very Idea of a Conceptual Scheme'. Here Davidson offers some
strong statements of the case for a truth-conditional semantics that
would *not* thus reduce to whatever is pragmatically acceptable or
(currently and contingently) 'good in the way of belief'. Nevertheless it has proved possible for interpreters like Rorty to latch onto
other, more concessive or ambivalent passages, those that would
ultimately relativize 'truth' to the entire set of truth-claims (or
communally warranted beliefs) which happen to prevail from time
to time within this or that cultural-linguistic context of enquiry.
And from here it is no great distance to the ultimately pared-down
(minimalist-semantic) argument that Davidson proffers – I think
for lack of any better, more adequate response – in 'A Nice Derangement of Epitaphs'. For as we have seen that essay retreats to a
point where the truth-conditions for speaker's intent and communicative uptake are defined very nearly out of existence. That is
to say, they are treated as products of a 'passing theory' whose field
of application is 'vanishingly small', whose devising is a matter of

ad hoc adjustment to this or that one-off speech-situation, and whose nature is, accordingly, to self-destruct – or pass into the limbo of (largely redundant) 'prior theories' – as soon as their task has been performed. At this stage there is really no difference – no difference that makes any difference, as the pragmatist would say – between Davidson's minimalized conception of 'truth' and 'theory' and the Rortian idea that such talk can be reduced without remainder to the currency of in-place consensus belief. For if hunches and more or less inspired guesswork are all we have to go on in interpreting speaker's meaning – and if truth *just is* whatever we impute to this or that utterance in context – then there seems little hope for any truth-based theory of meaning such as Davidson avowedly sought to provide in 'On the Very Idea of a Conceptual Scheme'.

What is thus thrown away is Davidson's entire case for rejecting those doctrines of cultural-linguistic relativism which traded on the notion of manifold and diverse 'conceptual schemes' – paradigms, language-games, discourses, ontological frameworks, and so forth – to make their point about the impossibility of assured translation between one and another such scheme. This whole line of thinking could best be overcome, as he then argued, by putting the emphasis back where it belonged: on the range of jointly logico-semantic resources which yielded a workable truth-theory for any conceivable (functionally adequate) language. But on the view that he adopts in 'A Nice Derangement' there is simply no room for such an argument. Interpretation becomes just a matter of feeling one's way around in a series of vaguely defined communicative contexts where syntax and semantics (even 'language' itself) have little to contribute and where understanding occurs, if at all, through the picking-up of various behavioural cues and clues that do not necessarily have anything to do with speaker's meaning or listener's interpretative competence. Hence the strange inversion of normative values by which malapropism figures as a test-case sample of what goes on in other (as we take it, more standard or commonplace) instances of verbal interaction. Hence also the Davidsonian idea of metaphor as just another, somewhat less quirkish species of malapropism that possesses no 'meaning', that doesn't belong to 'language', and which therefore – necessarily – defies all attempts to specify or theorize its structural-semantic attributes. It is an odd picture of how language works, and none the less so for Davidson's way of suggesting that this is how language-users just do get along, aside from all that off-the-point philosophical talk about truth, meaning, validity-conditions, and so forth.

VERSIONS OF TRUTH: DAVIDSON, DUMMETT, TARSKI

David Papineau pinpoints the problem with Davidson's minimalist semantic approach when he suggests just what it would take for a Tarskian theory to provide a substantive and useful (as opposed to a formalized but redundant) account of meaning and truth. 'The hard part of constructing T-theories', he writes,

> is to do with non-finite languages, languages where there are an indefinitely large number of sentences. If you were dealing with a finite language, you could straightforwardly, even if tediously and unilluminatingly, list the requisite T-theorems one by one. A large but finite set of axioms, which took the *s*'s one by one and simply stated the truth conditions of each, would still be a T-theory which specified the truth conditions of all the sentences of the language. But of course all serious languages are non-finite. . . . And so the tedious list-strategy will not be available in most cases. Tarski dealt with this by focusing on the internal *structure* of sentences. Even if there are an infinite number of sentences in a language, there will only be a finite number of words (dictionaries come to an end) and a finite number of ways of combining them. The infinitariness of the language will be entirely due to the fact that many of the ways of combining words can be iterated indefinitely. . . . And so it is in general possible to construct T-theories for infinite languages by discerning structure in complex sentences, and specifying the truth conditions of those sentences as functions of that structure. More precisely, one can state a finite number of axioms for the words, and the ways of combining words, which specify what contributions such components make to the truth of any complex sentences they occur in.[55]

In Davidson's version the Tarskian theory operates only as a source of assurance that some such formalized schema is always at hand – no matter how trivial or redundant – in order to define 'truth-in-L' for any given sentence of any given language, and thus to provide a ready defence against loose talk of 'conceptual schemes', 'ontological relativity' and the like. But this defence is quite useless, as

Papineau brings out, if it fails to specify *first* how these truth-conditions apply across the vast (potentially infinite) range of well-formed meaningful sentences, and *second*, how the truth-value of particular sentences relates to their 'internal' (i.e., syntactic or logico-semantic) structure. This is why Davidson's (apparently) strong line of truth-talk in 'The Very Idea of a Conceptual Scheme' gives way so easily to Rorty's suggestion that such talk be cashed out in holistic or pragmatist terms as simply what is 'good in the way of belief'. It is also why Davidson himself went on, in 'A Nice Derangement', to minimize the content of the Tarskian theory to a point of near-total redundancy. For at this stage nothing remains of it save an all-purpose 'principle of charity' according to which truth-values are in some sense logically prior to meaning, but where truth (or the attitude of 'holding-true') *just is* whatever we impute to speakers on the basis of some more or less plausible hunch.

I should admit that I once (not so long ago) took a different, more positive view of Davidson's accomplishment and for that reason argued – with at least some wavering support from Davidson himself – that Rorty had got him flat wrong.[56] For the record I still think that Rorty is wrong about the central issues, in particular his view that those same issues (of truth, meaning, valid interpretation, etc.) are so many tedious distractions from philosophy's scaled-down participant role in the 'cultural conversation of mankind'. But I should now be less confident in denying that Rorty can make out a plausible (albeit a partial and selective) case for his recruitment of Davidson as a half-willing ally in the postmodern-pragmatist cause. At any rate it meets with no effective resistance from the fall-back position that Davidson adopts in 'A Nice Derangement of Epitaphs'. For it is in the very nature of 'passing theories', as Davidson describes them, to be always underspecified with regard to their operative truth-conditions, and thus to fit in with the pragmatist idea that truth *just is* a matter of adjusting to localized (presumptively shared) habits of belief. This also follows from the 'principle of charity', at least as Davidson interprets it. That principle holds that we can only make a start in understanding what other people say and think on condition that we count them 'right in most matters', rather than supposing that they (or we) might be 'massively mistaken' on a whole range of basic issues, or possessed of such different (incommensurable) belief-systems as to render communication impossible.[57] Davidson offers this argument – in

'On the Very Idea of a Conceptual Scheme' – by way of rejecting those varieties of framework-relativism (Quinean, Kuhnian, Whorfian, etc.) which tend to endorse such a sceptical view. But of course there is a crucial difference – and one that has been the subject of much debate from Plato's *Theaetetus* to the present – between *belief* (or the attitude of holding-true) and *knowledge* as assessed by more stringent (non-speaker-relative) criteria of warranted or justified true belief.[58] In the end Davidson leaves himself with no choice but to duck this issue and identify truth with whatever counts as such according to our current interpretive lights.

Hence – as I have remarked – the symptomatic pattern of retreat which can be seen most clearly in his recent lecture-series on 'The Structure and Content of Truth'. Starting out with the call for a stronger (more substantive) theory than anything provided by the Tarskian schema in its generalized or abstract form, Davidson ends up – one suspects *faute de mieux* – by identifying truth with whatever sorts of attitude (or assenting disposition) we can reasonably impute to this or that speaker in this or that context of utterance. This is also to deny the possibility of our ever being in a position to conclude *either* that they (the speaker) had been wrong in entertaining some particular item of belief, *or* that we had misinterpreted their words on account of our either not sharing that belief, or sharing it on what later – with improved understanding – turned out to be mistaken grounds. In short, Davidson's 'principle of charity' is not – as might appear – a powerful argument for regarding truth (in any non-trivial sense of that term) as logically prior to issues of meaning and interpretation, and hence as playing a crucial role in his case against cultural-linguistic relativism. For what it really comes down to (as Rorty was quick to observe) is a redundancy-theory where truth-predicates are no sooner devised for all sentences in a language than they cancel straight through to leave those sentences in place with no better warrant than their hanging together with the rest of a speaker's naturalized beliefs or assenting dispositions.[59]

This is also why Davidson's theory – again contrary to first impressions – can so readily be enlisted in support of anti-realist doctrines which deny the existence of any truths beyond our current best notions of adequate descriptive, evidential, or explanatory warrant. These doctrines range all the way from Rorty's laid-back neopragmatist position to Feyerabend's self-professed 'anarchist' philosophy of science and – in most respects at the opposite extreme – Michael Dummett's analytically scrupulous argument that we *cannot make sense* of the idea of a 'reality' that would somehow transcend – or be at variance with – those current best notions.[60] Dummett has his own disagreements with Davidson (which I shall not go into here), while Rorty as we have seen makes selective but not altogether implausible use of Davidson's ideas. What they all three share is the anti-realist view which denies the possibility of even *conceiving* that there might exist truths (or facts of the matter) that are 'evidence-transcendent' (Dummett's phrase) or which cannot be known, represented or stated in our current, best-available terms. This fits well enough with the Davidsonian 'principle of charity', since here also the way appears blocked to any idea of truth which would question the baseline premise that we (and other people, other languages or cultures) must necessarily be counted 'right in most matters' if our and their beliefs are to make any sense.

As I say, Dummett is more aware than Rorty of the difficulties to which this argument can lead. They are especially acute for philosophy of science since here, more than anywhere, it is hard to deny both that progress has occurred in various scientific fields, and moreover, that such progress has often come about precisely through perceiving the limits of previous (in their time well-established) truth-claims, methods, and evidential grounds.[61] Of course it may be urged – from the relativist viewpoint – that this merely goes to show how 'progress' is itself just an honorific term attached to some favoured narrative account on which 'science' and 'reason' triumph over the forces of ignorance, irrationalism, and prejudice. Such is Feyerabend's standard debunking line and Rorty's more relaxed or accommodating notion of science as just another voice in the ongoing cultural conversation of mankind.[62] It is an argument which can also be derived – albeit in a less extreme, more qualified form – from Kuhn's ideas of meaning-variance and paradigm-change (especially during periods of 'revolutionary' science) as preventing any comparison between theories in point of their truth-content, their empirical adequacy, descriptive power, or proven explanatory yield. These relativist doctrines have been criticized from various quarters, and their validity questioned – as I think, convincingly – on each of the above (semantic, empirical, descriptive, and causal-explanatory) grounds.[63] However, my main point here is that Davidson's principle of charity in the end turns out to be perfectly compatible with a belief-based (speaker- and interpreter-relativized) theory of 'truth' which could accommodate any number of false, unwarranted, or downright irrational beliefs just so long as they conjured an assenting disposition from both parties concerned.

This is why Davidson is so easily coopted by the Rortian (blanket anti-realist) argument which counts all his truth-talk as just a harmless if somewhat tedious distraction on the road to a full-fledged pragmatist outlook. It is also why other commentators – Malpas among them – have looked to Heidegger and depth-hermeneutics as a means of fleshing out Davidson's account, or providing some significant (non-trivial) content for what otherwise appears a thoroughly vacuous theory. But these are desperate remedies indeed if one takes it on Davidson's (albeit somewhat faltering) assurance that the principle of charity has truth-preserving virtues beyond the mere appeal to consensus-belief or a pragmatist line of least resistance. And nothing could be further from his thinking than the kinds of vapid pseudo-profundity that result from this 'post-analytic' (or depth-ontological) turn against reason, logic, and any form of scientific or epistemological realism. For on one point at least Davidson is clear, both in his writings on truth, meaning and interpretation and in those brought together in the companion volume *Essays on Actions and Events*.[64] This is the conviction – inherited from Quine – that any adequate address to such issues will need to incorporate a physicalist (causal-explanatory) account which so far as possible avoids the resort to some deep further realm of meanings and motives beyond reach of rational analysis. In short, there is an element of interpretive violence – a violence (one might add) not infrequently encountered in Heideggerian readings – when these dilemmas and dubieties of Davidson's work are pushed in a depth-hermeneutical direction so markedly at odds with the general cast of his thought.

No doubt Rorty has a point when he remarks that the Tarskian schema, as Davidson deploys it, is 'a transcendental argument to end all transcendental arguments', or – what amounts to much the same thing – a truth-theory on which 'truth' drops out for all practical (non-trivial) purposes.[65] But this is by no means the end of the road, whether with regard to Tarski's theory, to Davidson's version of that theory, or again, to those aspects of his own (Davidson's) thinking which may yet be capable of other, more cogent interpretations. In fact it is one mark of his distance from Heidegger-inspired notions of the 'hermeneutic circle' and so forth that

Davidson's work clearly offers itself for further criticism and development of a rational-constructive kind. When approached in this spirit perhaps the oddest aspect of Davidson's case is his failure to pursue those alternative, more promising lines of argument that point a way beyond the dilemmas of anti-realism.

Concerning Tarski there is much disagreement as to whether his truth-conditional semantics should be seen as strictly neutral with respect to any further (ontological or epistemological) commitments, or whether – on the contrary – it must entail some argument for or against realisms of either variety.[66] His best-known statement on the topic might appear to count strongly against the consequentialist view. Thus: 'we may remain naive realists, critical realists or idealists, empiricists or metaphysicians – whatever we were before. The semantic conception is completely neutral toward all these issues'.[67] But there is also a strong argument – proposed by Richard Kirkham in his recent book *Theories of Truth* – to the effect that this remark can best be construed in the context of those narrowly positivist (or phenomenalist) theories that were prevalent in philosophy of science at the time.[68] In which case – and on other evidence from his writings – the remark will bear a different interpretation, one more in line with what Kirkham calls the 'default' attitude of commonsense ontological realism. Tarski would then be responding to his critics:

> If by 'naive' or 'uncritical' realism you are referring to the theory of justification or perception known by that name, you have misunderstood, for my theory is neutral about epistemological issues. But if you are referring to an ontological doctrine by these terms, you are right to attribute it to me, but you are only resorting to name-calling instead of making a genuine objection.[69]

Thus Kirkham argues first for the weaker thesis that nothing in Tarski's theory is incompatible with a realist ontology; second, for the somewhat stronger set of claims that 'even if ontologically neutral, it is not compatible with every possible non-realist theory of truth and may not even be compatible with every actual, every possible plausible, or every actual plausible non-realist theory of truth'.[70] And beyond that there are, he thinks, reasons to suppose that the theory may actually make better sense – avoid the charge of empty circularity and yield more genuine (substantive) results – if given a realist interpretation. In more technical terms:

the default interpretation of any philosopher who utters a declarative clause (such as the second conjunct of the basis clauses of
Tarski's definition or the right-hand side of the T-sentence) is to
take him or her as referring to a mind-independent world unless
and until we are given sufficient reason not to do so.[71]

On these grounds a case can be made for regarding both Tarski *and*
Davidson as committed 'by default' to a realist ontology in the
absence of which their arguments would be rendered very largely
trivial or redundant. That is to say, we should be justified in taking
Davidson at his word when he states – at the outset of his lecture-
series on 'The Structure and Content of Truth' – that the Tarskian
theory both *allows and requires* such a further realist component,
since otherwise it lacks (in his own terms) both substantive 'content' and any kind of logico-semantic 'structure' adequate to the
task in hand.[72]

Moreover, this suggests that we should *not* attach so much weight
to those subsequent passages where Davidson appears to back
down from his opening statement, or to hedge it around with such
doubts and qualifications that the lectures end up – as I noted
above – by retreating to an anti-realist position where truth just is
what we take it to be according to this or that (interpreter-relative)
set of meanings and beliefs. In other words the Davidsonian principle of charity needs strengthening to the point where it becomes
something more like a principle of rational reconstruction. For on
this reading his argument comes out both truer to his own best
insights and truer to those aspects of the Tarskian theory that best
make sense in terms of their shared commitment to a 'default'
(presumptively realist) ontology and epistemology. On the alternative, anti-realist account there is nothing in the nature of Davidson's argument (nor, for that matter, of Tarski's) which could
prevent the slide – or as Rorty would have it the wished-for *dénouement* – that issues in a pragmatist doctrine of truth as nothing more
than culturally warranted belief. In Kirkham's words:

> The fact that there can be a Realist, and hence nonvacuous, Tarski-
> like theory of truth has considerable significance, for the contrary
> assumption is a plank in the platform of a number of recent
> philosophical movements. Nonrealist answers to the metaphysi-
> cal question have enjoyed renewed popularity in the last few
> years partly because of the perception that there is no correct

nonvacuous alternative. Others have contended that the meta-physical project is itself ill-conceived in some way because there is *no* answer to it (not even a Nonrealist one) that is both correct and nonvacuous. . . . [T]he contention that there cannot be a non-vacuous Tarski-like theory of truth is a key premise in Rorty's argument that we ought to abandon philosophy in favour of a 'post-philosophical culture'.[73]

Thus a good deal hangs on this seemingly technical issue as to whether Tarski's theory is neutral between competing (realist and anti-realist) viewpoints, and whether, if not, it can be shown to favour (possibly even to require) some particular kind of realist ontology. I think that Kirkham is right in his suggestion that Tarski's more neutral-sounding statements were a product of his non-committal attitude *vis-à-vis* the sorts of philosophic doctrine ('naive realist', empiricist, positivist, phenomenalist, etc.) current at that time. Moreover the ground of debate has now shifted appreciably owing to the advent of causal-realist theories in philosophy of science and language. For it is a virtue of these theories that they allow a more adequate (certainly non-trivial) account of what it means for terms or statements to have reference, for truth-claims to possess substantive content, and for knowledge to consist in something more – something deeper – than the mere confirmation of existing beliefs through a (formalized or pragmatist) process of circular definition.[74]

Of course such forms of 'depth-ontological' enquiry have nothing whatsoever in common with the other (Heideggerian or depth-her-meneutic) usage of that phrase. Rather, they involve the basic sup-position that there exists a real world of mind-independent objects, processes, and events to which science gains access through a knowledge of their underlying (causal, micro-structural, or dispo-sitional) attributes. Discoveries of this type have explanatory 'depth' in virtue of their actually finding out more about the nature of that physical reality and developing more adequate theories by which to account for it. Nothing could be further from Heidegger's fixed antipathy to science, technology and all their works, that is to say, his notion of the scientific worldview as the terminal phase in a bankrupt epistemological tradition whose epoch had been marked – from Plato to Husserl – by the false dichotomy of subject and object and the growth of a narrowly instrumental (means-end) rationality.[75]

This is not to deny that certain aspects of Davidson's thinking (as indeed of Quine's) lend credence to the idea of a growing *rapprochement* between the 'post-analytic' and depth-hermeneutical projects. Among them – as I have said – are the Quine–Davidson turn toward holistic theories of meaning and truth; the thesis of onto-logical relativity which Quine takes fully on board while Davidson typically hedges his bets; and the occasional gestures in Davidson's writing toward a 'deeper' (more substantive) conception of truth which none the less remains so vaguely specified as to open the door to Heideggerian readings of the kind here described. But these apparent liabilities are better explained – as Kirkham persuasively argues – by noting their background in an earlier (empiricist, phe-nomenalist or 'naive realist') set of ideas about the scope and limits of scientific knowledge. As those ideas came under pressure from several quarters – from Quine's assault on the analytic/synthetic distinction, from Goodman's 'new riddle' of induction, from objec-tions to the covering-law (deductive–nomological) approach – so there developed that trend toward varieties of more or less extreme anti-realist reaction which I have surveyed in the course of this chapter.[76] However, there is now a large and impressive body of work in various fields (philosophical semantics, cognitive psycho-logy, history and philosophy of science) which puts the case for a causal-realist epistemology subject to none of those well-worn criti-cisms. Thus the prospects seem good for a working alliance be-tween, on the one hand, the kinds of depth-ontological explanatory approach essayed by philosophers of science such as Wesley Salmon and Roy Bhaskar, and on the other the causal theory of naming, necessity and natural kinds developed by Saul Kripke and Hilary Putnam.[77] (That the latter has now backed down from this argument in favour of an 'internal realist' position which ends up sounding much like Rortian pragmatism under a different name is no more reason in his case than in Davidson's to follow suit and count the argument a failure.)[78] At any rate realism is currently very much a live option and one that can claim a large measure of support not only from well-informed sources in philosophy of science but also from those theories of truth, meaning and interpre-tation which are most closely attuned to such thinking.

2

Complex Words *versus* Minimalist Semantics: Empson and Davidson

EMPSON ON TRUTH, MEANING, AND INTERPRETATION

In this chapter I shall put the case that William Empson's *The Structure of Complex Words* (1951) is among the most important and distinctive contributions to issues in present-day linguistics, literary theory, and philosophical semantics.[1] That the book has been largely ignored by philosophers – and received far less than its due share of attention from literary critics – is no doubt the result of its cutting across these conventional divisions of academic labour. Thus the critics have mostly been repelled or mystified by Empson's theoretical chapters (his apparatus of logico-semantic 'machinery') while the philosophers, Donald Davidson among them, have picked up one or two useful ideas but pretty much ignored the way that these ideas are put to work in the close-reading of texts.[2] This seems to me an unfortunate state of affairs since Complex Words is, among other things, the single most sustained and resourceful attempt to bring those disciplines together. In particular it offers a promising escape-route from the dead-end (as I see it) of Davidsonian 'minimalist semantics', that is, the idea that speakers and interpreters can simply get along on 'wit, luck, and wisdom' or with the aid of 'passing theories' that really come down to just intuitive guesswork plus an all-purpose optimizing 'principle of charity'.[3] Such arguments have lately exerted great appeal among post-analytical philosophers and also among a growing number of literary critics in the neopragmatist (or 'against theory') camp.[4] In what follows I shall therefore address myself to both readerships and hope to overcome the various sorts of prejudice that have so far prevented an adequate assessment of Empson's remarkable book.

His main purpose in *Complex Words* is to offer some account of what goes on when we encounter certain uses of language –

whether in poems, novels, or everyday speech – that are felt to
carry implications beyond their normal semantic range. Most often
there is a sense, on the reader's or listener's part, that the whole
weight of argument is condensed into some particular word in
context, thus allowing that word to communicate a kind of 'compacted doctrine', an order of implied entailment-relations between
two or more of its possible meanings. Thus Empson devotes the
early (and most densely theoretical) chapters of his book to a work-
ing-out of the various 'equations' – the structures of logico-semantic entailment – which explain how it is that language can achieve
such a variety of subtly suggestive effects. These are all derivatives
of the basic structure 'A = B', that is, the standard form of subject-
predicate logic which has dominated Western thinking from Aristotle to its latter-day revision and refinement by logicians such as
Frege and Quine. On the whole Empson accepts this model as
adequate for his own purposes, sometimes alluding to recent work
in the field (chiefly by Russell and F.P. Ramsey), but otherwise
deploying a fairly traditional range of analytic terms. Where his
book does break new ground – and where it should be of interest to
philosophers – is in its further claim that such modes of logical
analysis can usefully be carried over from the level of sentences,
statements, or propositions to that of individual 'complex words'.
By analysing the various orders of 'equation' or 'compacted statement' carried by certain words in context, the interpreter may
avoid having recourse to ill-defined rhetorical terms like 'ambiguity', 'paradox', or 'emotive' meaning. For the effect of such recourse – in literary criticism as in ethics, linguistics, philosophical
semantics and elsewhere – is to close off any prospect of improved
communicative grasp and thus make language appear more illogi-
cal, more prone to forms of irrational mystery-mongering, than is
actually (or typically) the case.

This is why, as Empson remarks in a footnote, 'the term Ambiguity, which I used in a book title and as a kind of slogan, implying
that the reader is left in doubt between two readings, is more or less
superseded by the idea of a double meaning which is intended to
be fitted into a definite structure'. (*Complex Words*, p. 103n; the
reference is to Empson's earlier and much better-known book *Seven
Types of Ambiguity*.[5]) Granted there are cases – 'Type IV' equations
as he calls them – where it is hard for analysis to get a hold since the
implied doctrine is either so vague (as in certain passages of Wordsworth) or so downright and flatly paradoxical (as in various state-

ments of religious faith) that it simply defies rational under-standing.[6] But even here, Empson argues, we still do better to assume that 'the human mind is not irredeemably lunatic, and cannot be made so' (*Complex Words*, p. 83n). Which is also to say that we can only make sense of such cases – figure out the deviant 'logic' involved – by seeing how they manage to *exploit or circum-vent* the more usual (rationally accountable) forms of logico-seman-tic equation. For it is against this background of everyday linguistic competence – our ability to assign meanings and motives on a basis of shared understanding – that we can at least grasp the rhetorical gist of such deviant ('Type IV') specimens. Moreover, Empson suggests, '[i]t is because the historical background is so rich and still so much alive . . . that one can fairly do what seems absurdly unhistorical, make a set of equations from first principles'. (p. 269) In other words, there is no contradiction – despite some degree of theoretical strain – between Empson's interpreting his various authors always with the *Oxford English Dictionary* to hand (that is, his concern to work within the range of plausible, historically documented meanings) and on the other hand his offering a gener-alized theory, a logical grammar of complex words with strong universalist claims.

This should really not appear so much of a paradox to anyone familiar (say) with Chomskian linguistics or related developments in cognitive psychology and philosophical semantics.[7] It requires nothing more than the working premise that speakers can produce (and interpreters understand) a vast, potentially infinite range of utterances on the basis of a finite – hence intelligible – stock of grammatical ground-rules, transformational structures, logical en-tailment-relations, and so forth. Indeed I have put the case else-where that Empson's approach – or something very like it – is the most promising (if so far neglected) candidate for a theory of se-mantic interpretation that would meet the requirements laid down by Chomsky in his later writings on the subject.[8] However, mym-ain concern here is with Davidson and his failure to pursue some of these paths opened up by Empson's work. Most important is his distinction between 'head meaning' and 'chief meaning', which I think Davidson must have had in mind when formulating his own (less elaborate) ideas about prior and passing theories.[9] For Empson, the 'head meaning' of a complex word is that which 'holds a more or less permanent position as the first one in its structure' (*Complex Words*, p. 38). There may well be various criteria for deciding on this, such as its 'being the most frequent in use or the one supported by derivation'. Moreover – and crucially for Empson's later chapters of applied literary analysis – 'a word may change from having one meaning as the head to having another, or a writer may impose a head meaning of his own'. (p. 38) Nevertheless it is the case – as with Davidson's 'prior theory' – that we are here in the region of a relatively generalized (high-level) order of semantic grasp where communicative uptake depends much more on the interpreter's being 'at home' in the language than on his or her responding to local indications that the word has undergone some semantic shift under pressure of context or speaker's/author's intent. Perhaps, Empson concedes, 'the term is merely a convenience which needs subdividing'; still 'it never refers only to the example of a use of the word which is being considered at the moment' (p. 38). It thus remains distinct from what Empson designates the 'chief meaning' in any given case: that which 'the user feels to be the first one in play at the moment', or which 'the speaker if challenged would normally pick out . . . as "what he really meant" ' (ibid.). And again, by way of helpful analogy: 'if the "chief meaning" is allowed a suggestion of local or tribal chieftains I think it can easily be remembered as applying only to a local occasion' (ibid.).

Empson distinguishes various ways in which a single word can 'carry a doctrine' or be felt to communicate some item of purportedly veridical belief. The first – and most fundamental – is the 'Existence Assertion', that which must standardly be taken to imply that 'what the word names is really there and worth naming'. Of course such assertions may be false, misleading, or fictitious, a point to which Empson repeatedly returns when discussing the more dubious varieties of 'Type IV' equation. For that matter, 'Aristotle pointed out that a syllogism might be regarded as a redefinition followed by a tautology', and something similar might apply to 'the word "electron" in the writings of any one physicist who claimed to put forward a complete atomic theory; all the properties should somehow be included in the idea' (p. 40). But one can best start out, Empson thinks, by adopting some basic formal notation (such as the existential quantifier $\exists$) which can then most often be taken for granted as a part of the implied background whenever some utterance is felt to carry any kind of assertoric force. This amounts to something very like Davidson's 'principle of charity', that is, the principle that speakers will normally mean what they say, that their meanings can only be construed on a basis of imputed (presumptively true) beliefs, and therefore that we must count them 'right on most matters' if they (and we) are to have any chance of achieving communicative uptake.[10]

The same rule applies – as both Davidson and Empson suggest – to cases where the belief in question may strike the listener/reader as either plain false or in need of further (perhaps more 'charitable') interpretation. Thus, in Empson's words, there is often the sense of an obscure truth-claim 'when really the mental goings-on are more confused'. Even so, 'a "verbal fiction" may need a great deal of looking into, and the full analysis of it may be very complex, but so far as the speaker is wholly deceived by it I think we must accept his own view that he is simply using it for an existence assertion' (*Complex Words*, p. 40). For otherwise – and up to this point he agrees with Davidson – we cannot make a start in understanding what they say on the basis of belief-ascriptions which may diverge widely from our own ideas of rationally warranted belief, yet which none the less enable us to figure out the speaker's gist by a process of (more or less complex) rational reconstruction. In short, we need to count them 'right on most matters' – or wrong only with regard to certain specific items of belief – if we are to see how some localized instance of mistaken reasoning or false analogy has got in the way of our construing their words at face (truth-preserving) value.

So it is, Empson argues, that we are able to interpret even the more irrational ('Type IV') equations as the upshot of obscure yet intelligible processes of thought which can provide a basis for communicative uptake despite all the deep-laid obstacles involved. However, as I have said, there is a crucial difference between Davidson's and Empson's way of applying this generalized principle of charity. For in Empson's case it requires a recognition that some beliefs are indeed false, and that to interpret them properly – that is, against the background of warranted or justified belief – is to see just how far, and in what precise ways, they deviate from the kinds of logico-semantic structure that characterize our 'normal or waking habits of thought'. On Davidson's account, conversely, the principle assumes such a blanket and all-purpose 'charitable' guise that it leaves no room for elementary distinctions between truth and the attitude of holding-true, or veridical belief and whatever lays claim to that title on the say-so of this or that individual speaker, Quinean 'native informant', or local community of like-minded persons.[11]

Newton-Smith makes this point most effectively when he remarks – à *propos* various relativist trends in present-day philosophy
of science – that there is something very wrong with a theory where
every belief necessarily comes out right just so long as we apply the
standard Davidsonian rule. Thus: 'it is fashionable to argue in some
quarters that one can transcendentally justify a principle of charity
which ordains us to endeavour to maximize the ascription of true
beliefs in the interpretation of the discourse of others'.[12] But the
effect of this ordinance – if carried right through – is to undermine
the very distinction between truth and falsehood by confusing
what speakers or thinkers may *truly be held to believe* with what they
rightly or justifiably believe according to the best available criteria of
theoretical consistency, valid argument, evidential warrant, and
so forth. Quite simply, 'we do not always want to be charitable',
and that for reasons which may indeed dispose us to count other
people wrong on some matters, but which thereby promote the
shared human interest in advancing knowledge through forms of
rational, constructive, truth-seeking endeavour. As Newton-Smith
puts it:

> [w]hatever plausibility the claim has that the principle of charity
> is a sort of *a priori* constraint on the interpretation of the ordinary
> discourse of others, it has no plausibility as a constraint on the
> theoretical discourse of others. The simple reason is that we well
> understand how easy it is to have a theory that turns out to be
> totally incorrect. While it may be hard to see how a group of
> people could cope with the everyday world in the face of massive-
> ly mistaken beliefs (this is what gives the principle its plausi-
> bility), it is easy to see how a group can be utterly mistaken at the
> theoretical level.[13]

Thus Davidsonian 'charity' comes out looking more like a vote of
no confidence in human reason, or – what amounts to much the
same thing – an approach that allows for no adequate distinction
between knowledge arrived at through a process of disciplined,
self-critical enquiry and beliefs (or attitudes of holding-true) whose
warrant is a matter of their happening to fit with other such currently accepted items of belief. Thus a way is left open for cultural
relativists or postmodern-pragmatists like Richard Rorty to recruit
Davidson in support of the idea that truth *just is* whatever counts
as such within this or that interpretive community.[14]

What is needed in order to avoid this unfortunate upshot is a theory of truth (or of validity in interpretation) which conserves the principle of charity up to a point but which doesn't push that principle so far that it collapses into a kind of all-licensing relativist outlook. This is why Empson is so careful to distinguish between, on the one hand, the idea that we should interpret speakers as most likely having reasons (or intelligible motives) for saying and believing what they do, and on the other the equally important principle that those beliefs may on occasion be wrong – products of false analogy and the like – and thus interpretable only by way of a more complex (critical-evaluative) treatment. Thus there are, on his reckoning, five basic 'ways in which a word can carry a doctrine'. The first has to do with existence-assertions – symbolized by Russell's $\exists$ – whose truth-conditions may well be obscure, or whose analysis may lead rather quickly into regions where there seems little choice but to accept, with Davidson, that 'truth' must be relativized to speaker's belief or the attitude of holding-true. At the limit, Empson thinks, 'the complexity of the word is simply that of the topic', so that (for instance) 'most newspaper headlines . . . must be supposed to make assertions of this sort' (*Complex Words*, p. 40). Such cases are therefore not of much interest for his own purpose, except in so far as 'the feeling [they give] of simplicity and irreducibility is often borrowed by the other types, to make themselves stronger, so that they are easily confused with it'. However, the prospects improve very markedly – so Empson maintains – when attention shifts to the logical 'grammar' (or the structures of logico-semantic implication) whose range is roughly covered by the other four ways in which doctrines are carried by words. For these are less open to the kinds of rhetorical imposition – by false analogy, concealed premises, affirming the consequent and so forth – which can easily work their effects where the whole weight of argument is borne by some strong if obscure use of the existence-assertion.

I had better now offer some sample passages from *Complex Words* in order to convey what Empson wants to do with his pieces of logico-semantic 'machinery'. The first two chapters describe the theoretical scope of his project and – I would argue – suggest just how far short of it Davidson falls by retreating to a minimalist-semantic position where there seems little point in maintaining the distinction between head and chief senses, 'prior' and 'passing' theories, or matters of generalized linguistic competence and matters of local interpretative grasp. First Empson's chart (p. 54) of the five equations according to the implied order of priority between subject and predicate:

The major sense of the word is the . . . Subject   Predicate

The sense demanded by the most
immediate context is the . . . Subject   II   I
. . . . Predicate   III   V

The order of the two senses is indifferent: IV

Type IV would seem to cover existence-assertions of the vaguer, more inclusive or logically recalcitrant kind, as well as 'deep' paradoxical truth-claims or instances (like Wordsworth's pantheistic 'sense') where 'the middle term is cut out' and where 'the whole poetical and philosophical effect comes from the violent junction of sensedata to the divine imagination given by love' (p. 296). But we can get more idea of how the method works from Empson's description of the various sorts of equation – or orders of logico-semantic entailment – that are carried by complex words. For it then becomes clear that such 'deviant' (Type IV) cases can best be understood against the normative background provided by Types I to III and V. And this applies even to those puzzling cases where it seems that the whole weight of argument is borne by some paradox or false analogy which must either be taken on its own rhetorical terms (as carrying an obscure but powerful existence-assertion) or else treated as merely 'emotive' and hence beyond reach of rational understanding. In so far as we are able to interpret such instances at all – rather than consign them to some private dimension of associative whimsy, chronic malapropism, linguistic psychopathology, or the like – they cannot be entirely discontinuous with what Empson calls our 'normal waking habits' of logico-semantic grasp. After all, as he says, 'I am trying to write linguistics not psychology; something *quite* unconscious and unintentional, even if the hearer catches it like an infection, is not part of an act of communication' (p. 31).

This may give some idea of how much has dropped out in the passage of translation from Empson's highly elaborated theory to Davidson's minimalist-semantic approach. Of course the comparison would have less point if Davidson had never read Empson or discovered nothing of interest in *Complex Words*; but there is evidence enough to suggest otherwise, both in footnotes and the general drift of his thinking between 'On the Very Idea of a Conceptual Scheme' and 'A Nice Derangement of Epitaphs'. Besides, it seems to me – to strengthen the case – that Davidson's work might have offered more resistance to the current neopragmatist/'against theory' trend had he attended more closely to Empson's passages of detailed logico-semantic exegesis. The point is best made in connection with those kinds of obscure existence-assertion that Empson thinks it important to analyse so that we are not too much at the mercy of irrational creeds or doctrines. Thus:

> I think that the same feeling of assertion is carried over to an entirely different case, which I shall call an 'equation' and propose to divide into four types. Two senses of the word are used at once, and also (which does not necessarily happen) there is an implied assertion that they naturally belong together, 'as the word itself proves' . . . . The most frequent are 'A is part of B', 'A entails B', and the more peculiar one 'A is typical of B' will have to be introduced. By definition an equation always generalizes, because if it only said '*This* A is B' the effect would only be a double use of the word (which I symbolize 'A.B'), imputing both 'A' and 'B' to 'this', and there would be no compacted doctrine. However it may presume a limited view of 'A', probably one with vague limits, and one could describe this process as saying 'A's *of this sort* are B'. When the sort is clear at the time of speaking the analyst ought to write instead of 'A' the narrower definition, but the speaker may be so vague that this would misrepresent him; for example the sort required may be recognised only as what suits the Emotion or Mood. In particular you can get equations of the form 'a normal A is B', or 'A good A is B', and I do not know that this last has any important difference from 'A ought to be B'. Here we get a case where 'A=B' could apparently be the same as 'B=A', using different interpretations for them; because 'a normal A is B' is the same sort of thing as 'B is typical of A'. I think however that the two orders have different effects, and so are worth distinguishing; the first can be taken very lightly, but the second makes you hold a specific doctrine which you are likely to remember. (*Complex Words*, pp. 40–1)

Empson's point, once again, is that we are able to interpret such deviant equations only by analogy with other, more normal or rationally accountable processes of thought, and then (in so far as this 'machinery' comes into play) that we can and should resist their more 'contagious' effects. Such resistance is especially called for, Empson thinks, in cases where the word is held to convey some authentic primordial wisdom or some order of deep paradoxical truth-claim beyond reach of mere analytical intelligence.

Thus Empson cites Archbishop Trench – one of the first editors of the *Oxford English Dictionary* – arguing for the doctrine of Original Sin by way of an appeal to the two senses ('pain' and 'punishment') supposedly conjoined in the Latin *poena* (*Complex Words*, pp. 81–2). And a footnote takes George Orwell more mildly to task for suggesting, in his novel *1984*, that human beings can be made to accept any number of irrational or mind-wrenching paradoxes (such as 'war is peace' and the other slogans of Newspeak) once the habit gets a hold through continued exposure. Empson doesn't deny that such cases occur and can sometimes exert a powerful influence through forms of religious or political indoctrination. Thus 'what he [Orwell] calls "double-think, a process of intentional but genuine self-deception, easy to reach but hard to hold permanently, really does seem a positive capacity of the human mind, so curious and so important in its effects that any theory in this field needs to reckon with it' (p. 83n). On the other hand, Orwell was perhaps too quick to generalize from the imaginary 'nightmare world' of *1984* to a gloomy conviction that people could always be swung into accepting any such irrational belief. 'The emotional ground of the process', Empson thinks, 'is a secret but fully justified fear'; all the same 'the case is so hideously special that it seems rather hard to generalize'. And again: 'you can have a usable linguistic theory which doesn't apply to sheer madness, just as you can have a wave theory which doesn't apply to cases of turbulence; but this is not much help unless you have some means of knowing where, and how often, turbulence is likely to occur' (ibid.).

This footnote is important because it shows Empson testing the limits of his theory against an example which might well appear to throw that theory into doubt. His conclusion has the kind of flatly commonsensical tone which may strike some readers as really little more than an evasion of the main issue, a refusal to acknowledge how powerful are the forces that make for unresisting acquiescence in these forms of irrational belief or wholesale paradox-mongering. Such is Empson's final declaration – *contra* Orwell – that 'the human mind, that is, the public human mind as expressed in a language, is not irredeemably lunatic and cannot be made so' (p. 83n). But the point of his taking *1984* as a 'hideously special' case is that it offers an extreme (hence untypical) example of one possible way that language or the human mind can go, a case where Type IV equations operate at full paradoxical stretch and where his theory of complex words encounters the greatest challenge to its powers of interpretative grasp. Still there is a sense in which the novel's whole effect comes from its exploiting possibilities of mass-induced linguistic and political 'double-think' which can only be perceived as such against a background of normative (logico-semantic) assumptions.

Thus the Orwell case is not so remote from those poetical uses of paradox or 'Type IV' equations – such as Keats's 'Beauty is truth, truth beauty' – where again 'the assertion goes both ways around' and may well strike the reader as a facile or unearned closing rhetorical flourish. Empson agrees that there is something rather wrong about a statement that bears such a weight of thematic implication while apparently ignoring all the obvious objections (for example, 'ugly truths' like the facts of disease and human suffering) which rise up against it as soon as we ask what the doctrine actually amounts to. Perhaps this is why so many people (Aldous Huxley and Robert Bridges among them) had found Keats's line so offensive: 'there is a flavour of Christian Science; they fear to wake up in fairyland, and probably the country of Uplift' (p. 373). Still it is no answer, Empson thinks, if we take one or other of the currently available lines of least resistance, such as I.A. Richards's talk of poetic 'pseudo-statements' or 'emotive' meaning, or again, Cleanth Brooks's idea of paradox as the structuring principle of all great poetry.[15] For the problem with both approaches is that they rescue Keats from his prosaically minded detractors only at the cost of ignoring his efforts to communicate something of importance in the poem's closing line. Thus one needs to work through all the possible 'equations' involved – the various orders of co-implicated statement carried by the words 'truth' and 'beauty' – before giving up on the attempt to make adequate sense of them. Even then we shall better understand what the poem is about – and not be tempted to dismiss the last line (with Bridges) as a piece of 'flashy' pseudo-philosophy – if we see how hard Keats has worked to bring it around to a conclusion which may not be justified on the strictest (logico-semantic) terms but which yet has considerable power in context. Thus '[t]here should not be a complacent acceptance either of "some indefensible sense" or of a merely emotive stimulus; the thought of the reader needs somehow to be in movement' (*Complex Words*, p. 371).

That is to say, we underestimate the intelligence of poet and reader alike if we take it that analysis has to stop short whenever it is confronted with some instance of paradox or (seemingly) 'emotive' language. The case is much the same for Keats's or Wordsworth's Type IV equations, for Orwell's examples of 'News-peak' or 'Doublethink', and also – I would argue – for those instances of chronic malapropism which Davidson offers as supporting evidence for his minimalist theory-to-end-all-theories.[16] That is, they each involve a more complex process of thought – on the speaker's, poet's, interpreter's, or theorist's part – than is anywhere provided for in Davidson's account or by literary-critical doctrines of paradox or emotive meaning. Empson offers various examples of this process at work, some of them – the more extended and elaborate – drawn from literary texts. They include the word 'fool' (or 'folly') as used by writers like Erasmus and Shakespeare with implications varying from 'wise or holy innocence' to cynical self-hatred and disgust; 'honest' as it figures in Renaissance and Restoration usage, again with a wide variety of positive and negative tonings; and 'dog' (like the kindred 'rogue') as a sort of mock-abusive, half-admiring, even affectionate slang-term: 'true to one's instincts (which cannot be all bad), hence trustworthy and reliable in a down-to-earth, doggish, un-self-deluding way'. The point about these words, as Empson interprets them, is that they possess a certain typical 'humour of mutuality', a reflexive or self-implicating tone which also goes along with a decent measure of other-regarding sentiment. This is not to deny that the process can be used to very different, ethically repugnant ends, as with 'dog' in the misanthropic tirades of *Timon of Athens*, 'fool' in some of Lear's more deranged utterances, or Iago's performance with the keyword 'honest' which he uses to snare Othello in the trap of a murderous paranoid delusion. But in these cases, as with *1984*, it can be argued that the whole 'dramatic and psychological effect' results from our sense of how the words have been somehow twisted away from their normal implications and contexts of usage.

To some extent this has to be regarded as an exercise in historical semantics, or a method which finds its preferred home-ground in those periods – like the early- to mid-eighteenth century – marked by the emergence of a secular-humanist ethos. Hence Empson's constant resort to 'that majestic object', the (Oxford) *New English Dictionary on Historical Principles*, as a richly documented source of examples and evidence. But this should not be taken as a limiting judgement on his claim to provide a generalized theory of interpretation, a method that extends beyond particular (period- or culture-specific) examples to the 'logical grammar' of complex words or the means by which communication comes about *despite and across* differences of culture, worldview, or interpretative ethos. For there is no reason – methodological prejudice aside – to suppose (with Davidson) that 'prior' theories' are pretty much useless for interpreting particular speech-acts in particular contexts, or conversely, that such instances can offer no support for anything beyond the most minimal (*ad hoc* or 'passing') theory of what goes on in the process of communicative uptake. Granted, '[the connection between theory and practice, where both are living and growing, need not be very tidy; they may work best where there is some mutual irritation' (*Complex Words*, p. 434). And of course it may be argued that most speakers and interpreters get along well enough without possessing a conscious ('theoretical') grasp of just how the process works. But it is a different matter when Davidson claims that any such project is radically misconceived since philosophy of language (or interpretation-theory) has nothing to explain beyond the mere fact that speakers and interpreters *just do* – for the most part – manage to communicate. For in that case one might as well argue that no theory, whether in the human or the natural sciences, can possibly do more than put itself out of business by confirming what everyone already ('preconsciously') knows as a matter of practical, commonsense, or intuitive grasp.

Not that Empson lacks all sympathy with this line of 'against-theory' argument. After all, as he remarks, '[the] whole notion of the scientist viewing language from outside and above is a fallacy: we would have no hope of dealing with the subject if we had not a rich obscure practical knowledge from which to extract the theoretical' (p. 438). However his target here is the behaviourist approach (represented chiefly by Leonard Bloomfield) which Empson sees as a misguided attempt to emulate the natural sciences in a different field of study, and to do so – moreover – on a narrowly empiricist view of how scientists think and work.[17] This was already Empson's argument in the closing chapter of *Seven Types of*

Ambiguity, where he makes the same point by way of an analogy with the current (post-1920) situation in particle physics and the quantum-theoretical domain.[18] But it is well to be clear just what Empson means by drawing this suggestive analogy. He is *not* putting forward an early version of the now fashionable postmodern-pragmatist line, that is, the idea of interpretation – whether in the human or the natural sciences – as observer-relative and hence open to all manner of diverse (incommensurable) claims.[19] Nor is he arguing that there exist no criteria of truth, validity or knowledge aside from those currently adopted by this or that 'interpretive community'.[20] Still less would Empson endorse the radical subjectivist view which exploits all the well-known 'paradoxes' of quantum theory (complementarity, the wave/particle dualism, the uncertainty-principle, the observationally induced 'collapse of the wave-packet', and so on) in order to maintain – in its crudest version – that physical reality *just is* whatever we make of it.[21] On the contrary, he argues: the point of having theories in the human or the natural sciences is precisely to make due allowance for that factor of interpretive or observational involvement which positivism failed to acknowledge. In neither case does this entail a retreat to the sorts of extreme relativist or subjectivist position which were already gaining among literary intellectuals – and some commentators on the new physics – when Empson first addressed these issues in *Seven Types of Ambiguity*.

The best way to understand such attitudes, he suggests, is to see them as a largely reactive or defensive response to the claims of logical positivism in its earliest, most doctrinaire form. For if indeed it is the case – as this programme held – that the only classes of meaningful statements are (1) those that can be verified by experiment or empirical observation, and (2) those whose truth is a function of their purely analytic (hence tautologous) logical structure, then it follows that all other statements – whether in ethics, literary criticism, or the vast majority of everyday social and communicative contexts – amount to no more than a species of 'emotive', 'metaphysical' or strictly meaningless talk.[22] But this doctrine soon ran up against problems, not least the fact that it proved incapable of being stated in a form that satisfied either of its own requirements, that is, the verification-principle or the condition of strict analyticity.[23] Moreover, it was something of an irony that the doctrine should be propounded by philosophers of science at just the time when advances in the quantum-theoretical field were making it increasingly difficult to maintain any such hard-line positivist account of language, truth, and logic. These considerations are all in the background, I think, when Empson puts forward his analogy between the current (late-twenties) situation in particle physics and the question how far we can interpret language – poetic or everyday language – as communicating truth-claims that don't come down to just a matter of 'emotive' pseudo-statement or (what amounts to much the same thing) subjective response on the interpreter's part.

This is why he takes issue with I.A. Richards's claim that 'the function of poetry is to call out an Attitude which is not dependent on any belief open to disproof by facts', and moreover that 'awareness of the nature of the world and the development of attitudes which will enable us to live in it finely are almost independent' (*Complex Words*, p. 7).[24] It is clear, Empson goes on, 'that *almost* might become important here', since it lets in at least the possibility that 'one needs more elaborate machinery to disentangle the Emotive from the Cognitive part of language' (ibid.). *Complex Words* in effect takes up this challenge and offers a full-scale apparatus of interpretive 'machinery' by which to explain how such vague talk of 'emotions' and 'attitudes' might be rendered more precise – more rationally, humanly, and ethically accountable – through a further stretch of logico-semantic analysis. Thus: '[m]uch of what appears to us as a "feeling" (as is obvious in the case of a complex metaphor) will in fact be quite an elaborate structure of related meanings' (pp. 56–7). And again: although 'Emotions and Moods may well be important in calling out and directing the interaction', still 'it seems clear that the first thing to examine is the result, what might be called the logic of these unnoticed propositions' (p. 57). For the trouble with emotivist doctrines, whether in literary criticism or ethics, is their willingness on the one hand to endorse the claims of logical positivism at face value, and on the other, to discount the truth-content of any language – everyday or poetic – that doesn't conform to those same (in any case impossibly restrictive) standards. All of which offers good grounds for supposing that 'the emotions in the words will normally evoke senses that correspond to them (except in swear-words, intensifiers, and for that matter raving), and the structure to be examined is that of the resultant senses' (p. 52). Otherwise – on the emotivist view – there would seem no limit to what 'language can get away with' by imposing false or irrational beliefs, like the paradoxes of Orwellian doublethink, which exploit the presumed incapacity of human thought to muster any defence against them.

It is worth quoting Empson once again to see what is involved in this (sometimes very complicated) process of 'putting emotions into an equation form'. With the simplest cases, he suggests,

> an underlining of the Stock Emotion 'A.A!' may be assertive, as if saying 'A does deserve the emotion commonly given to it', and we could write this 'A = A!'. But I think this can be classed as a kind of existence assertion. In the same way the denial of the stock Emotion 'A. − A!', as in 'jolly lust', can be viewed as a denial of the reality of the supposed 'thing'. Of course the function of 'jolly' is to act as a context which kills the stock emotion, and I agree that the first effect is this and nothing more; but we are at once driven to look for a new Sense of *lust*, whether successfully or not, which will account for such a use; and people who fail to find one would tend I think to call it a 'cynical denial that there is any such thing as lust'. In any case there seems no great advantage in extending the equation form to this sort of case. The important type would be 'A.B!', and I deny that an equation 'A = B!', without further complications of sense, ever occurs; at any rate without being a recognisable error of a kind very unlikely in a native speaker of the language. (*Complex Words*, p. 41)

Without (for the moment) pursuing these complexities any further we can now perhaps see what is involved in Empson's opposition to emotivist doctrines and his general case for this approach to language *via* a logico-semantic theory of complex words. More specifically – and more to the point for my argument – it suggests what is missing from Davidson's under-theorized treatment of related issues. For there is a sense in which Davidson makes the same error as Richards. That is, he assumes that any such theory will encounter all the problems that were faced at an earlier stage by logical empiricism in its failed attempt to demarcate the realms of meaningful and meaningless statement. Just as Richards fell back on an emotivist doctrine through being over-impressed by that programme, so Davidson retreats (in 'A Nice Derangement of Epitaphs') from 'prior' to 'passing' theories, and thence to a notion of communicative uptake in which 'theory' pretty much drops out and interpretation becomes largely a matter of *ad hoc* response to contextual cues and clues.

MAXIMIZING TRUTH: CHARITY OR HUMANITY?

Empson's is one promising path not pursued by Davidson and other exponents of the 'post-analytical' line. A second is the naturalized causal-realist approach as applied not only in philosophy of science but also (by writers such as David Papineau) to issues in epistemology, cognitive psychology and philosophical semantics.[25] Papineau is particularly clear and convincing on the problems with any purebred Tarskian theory, that is to say, any formalized semantical approach where meaning is construed as somehow dependent on truth-conditions but where 'truth' becomes merely a product of circular or abstract definition, a term that stands in for the lack of any other (more substantive) account.[26] Thus the upshot of Davidson's 'principle of charity', consistently applied, is to rule out any possibility that we might have warrant for counting other people wrong (deluded, misinformed, in the grip of some prejudice or other) and yet have correctly interpreted their gist on the basis of a justified belief-attribution. That is, it offers no workable account of how (1) we can properly interpret a speaker's meaning as given by the implicit truth-conditions for his or her utterance, while (2) at the same time thinking them wrong in holding that particular belief. For, as Davidson argues, there is simply no way that interpretation can get off the ground unless we suppose that speakers will standardly mean what they say and have adequate grounds (or veridical warrant) for speaking, thinking, and believing as they do. Otherwise we should run foul of the principle of charity which enjoins us to maximize the truth-content of their utterances on pain of failing – or wantonly refusing – to interpret their words as intended.[27]

However, there is something very odd about a theory that makes such 'charitable' (truth-preserving) claims while so blatantly flouting our best intuitions with regard to the difference between truth and falsehood. That is to say, it ignores the always-present possibility that speakers may *in fact* be wrong in their beliefs or subject to various kinds of error, confusion, or partial understanding. After all, as Papineau remarks, 'people do sometimes say false things', so that any theory (no matter how charitable) which excludes that possibility *a priori* is, on the face of it, not a strong candidate for truth-preservation. What it lacks most conspicuously is any means of attributing *false but intelligible* beliefs, such that we can grasp how and why a speaker might have come to hold them while declining – on our own part – to regard them as true or as adequately borne out by the best evidence to hand. This would be a 'principle of humanity' – Papineau's term – as opposed to that all-purpose Davidsonian principle of charity which effectively levels the difference between knowledge and belief, truth and falsehood, or the ascription of cogent validity-claims and the ascription of attitudes (like that of holding-true) which may not possess any genuine veridical warrant. Where 'humanity' comes in is through the process of our making due allowance for the range of possible causal-explanatory factors – limited knowledge, cultural prejudice, restricted access to relevant information-sources – which might help to make sense of why speakers say and believe what they do. Thus, in Papineau's words, 'the whole point of the switch to humanity is that on occasion we find it *intelligible* that *in such and such circumstances* people are likely to have *false* beliefs'.[28] For there is not much point in having a principle of charity whereby all sentences, meanings or belief-ascriptions necessarily come out true no matter how remote from our own best methods for distinguishing truth from falsehood.

Papineau argues this case most effectively as against Dummett's anti-realist conviction that there is not and could not be any such 'truth' beyond what is verifiable according to the standards of this or that epistemological framework.[29] 'The difference between my position and Dummett's', he writes,

> is that I have a more full-bloodedly realist notion of truth. My approach to truth allows that people can have certain beliefs, and manifest that they have them, without imposing any *a priori* requirement that their methods for acquiring such beliefs should lead them to the truth. Dummett, on the other hand, even when he is contemplating the possibility of 'verification-transcendent' truth conditions, still doesn't free himself from the idea that truth is *per se* what people's judgemental methods relate them to. So it is hardly surprising that he finds the idea of epistemologically inaccessible truth conditions incoherent.[30]

Thus the 'principle of humanity', as Papineau construes it, doesn't amount to an arrogant presumption that 'we're right and they're wrong', or a foolproof method for distributing the values of truth and falsehood so that we come out automatically in possession of the better, more cogent arguments. On the contrary: it is realist in the sense of entailing a symmetrical ('us and them') allowance first that *all* beliefs are in principle subject to correction; second, that our own (as well as their) best ideas of truth may prove false or inadequate; and third, that this recognition follows – *contra* Dummett – from a shared understanding that the way things stand in reality may not be captured by any currently available set of verification-procedures. Still we can have good grounds – epistemic, evidential or justificatory warrant – for attributing false beliefs to other people while none the less claiming (*contra* Davidson) to correctly interpret those same beliefs. What makes this possible is also what saves such an argument from the charge of mere arrogance or cultural imperialism brought against it by relativists of various stripe. For the requirement is not only that knowledge be advanced through, for example, the criticism of naturalized 'commonsense' thinking in proto-scientific cosmologies, or again, by offering better, more descriptively adequate or depth-explanatory theories in our conception of the micro-physical domain. What is also required – as in Papineau's 'principle of humanity' – is that the new theory should manage to explain just why other thinkers should have held (or should continue to hold) very different beliefs, and for reasons which come out as humanly intelligible (though not as true) on the proffered account.[31] For it is then possible to avoid both the relativist upshot of Davidson's argument and the opposite mistake of supposing that others can have had no reasons – no justification – for beliefs which we now (perhaps with good warrant) consider false or inadequate.

The most important point here is that this casual-explanatory model carries over from the realm of epistemology (how we and others get knowledge of the world) to that of belief-attribution where it is more a matter of figuring out the meanings, theories, or truth-conditions that obtain for certain speakers in certain, more or less well-defined contexts of enquiry. Papineau makes the point by way of an example: the fact that many 'naive' observers have the idea that they are viewing something green when in fact what is before them is a red object (like a London bus) under abnormal perceptual conditions (e.g., sodium lighting). Such people, he writes, are for the most part 'quite capable of discovering that sodium lighting is deceptive in this way . . . and then ceasing to form the relevant beliefs automatically in the relevant circumstances'.[32] And again, generalizing from this example: '[p]eople can reflect on the dispositions that give rise to their beliefs. Moreover, they can reflect on whether those habits are such as to give them true beliefs reliably. And then they can endorse or reject those habits accordingly.'[33] Thus the causal-explanatory approach is one that has decisive advantages for any theory of truth and interpretation that would seek to avoid the Davidsonian slide toward a Principle of Charity according to which we just can't understand what people say unless we count them necessarily 'right in most matters'. That is to say, it serves *both* as a means of reflecting critically on our own procedures of belief-acquisition *and* as a method for interpreting other people's beliefs in so far as they, like us, can be assumed to have arrived at those beliefs partly in consequence of various causal (or causally explicable) factors and partly through processes of analogy, inductive reasoning, inference to the best explanation, and so forth. In short, it follows Davidson in applying the Principle of Charity *so far as possible* in order to maximize the truth-content of their utterances and thus secure a basis for mutual comprehension. But it still leaves open the alternative possibility that some of their beliefs may come out false according to the best available criteria of reason, valid inference, or evidential warrant. In this case, Papineau argues, we should do much better to switch to the 'Principle of Humanity'. For then we shall recognize that others (like ourselves) are error-prone at least on occasion; that such errors come about most often for causally or otherwise accountable reasons; and hence that we *can* make a claim to have rightly interpreted their beliefs even if (as sometimes happens in our own experience also) those beliefs turn out to be erroneous.

I have suggested that the problems with Davidson's approach – his 'transcendental argument to end all transcendental arguments', as Rorty puts it – result from his adopting a highly restrictive view of what is involved in the process of understanding language on a truth-conditional (or belief-attributive) basis.[34] There are two main gaps in his theory, both of which, ironically, he appears to recognise but makes no serious effort to repair. One has to do with the causal component of meanings, attitudes, and beliefs, a topic very close to his interests in the volume *Essays on Actions and Events*, but one that figures hardly at all in his writings on truth and interpretation.[35] Indeed it could be argued – though I have no room to develop the argument here – that Davidson's idea of 'anomalous monism' as a means of coping with the mind/body or free-will/determinism issue is unsatisfactory (or evades that issue) in much the same way as his minimalist approach to semantics in 'A

Nice Derangement of Epitaphs'. That is, he leaves it very much a mystery how agents might yet have reasons to act (or speakers have meanings to communicate) despite the fact that their actions and beliefs are in principle subject to causal explanation and treatment in reductionist or physicalist terms.[36] What is required is a different kind of causal-explanatory account, one that allows for our ability to acquire more adequate, reliable beliefs through reflection on the various possibilities of error (perceptual illusion and so forth) that we and others have learned to recognize. For as Papineau remarks (*à propos* the sodium lighting example) 'we can ask, of any belief-forming habit, whether or not the beliefs it gives rise to are usually true'. And moreover we can feel quite safe in assuming that 'people will generally want to continue with just those beliefs that are reliable for truth, and eschew those which aren't'.[37]

Of course it may be said that Davidson adopts just such a causal-explanatory approach in the other main portion of his work to date, i.e. his *Essays on Actions and Events* and later writings on that topic. Indeed the most frequent criticism of this work has to do with its subordinating reasons to causes, or its treatment of actions as simply a subclass of events, to be explained according to causal regularities which leave no room for meaningful ascriptions of agency or purposive human intent. Thus Davidson's idea of 'anomalous monism' in fact works out as a systematic programme for excluding such (to him) problematical notions and hence reducing the philosophy of action to a matter of first-order causal explanation where intention figures in the merest of epiphenomenal or supervenient roles. As Paul Ricoeur puts it:

> Although [Davidson's theory] begins by stressing the distinctive teleological character of action among all other events, this descriptive feature is quickly subordinated to a *causal* conception of explanation . . . . In Davidson's strategy, causal explanation serves in its turn to insert actions into an ontology, not a hidden one but a declared ontology, which makes the notion of events, in the sense of incidental occurrences, a class of irreducible entities placed on an equal footing with substances in the sense of fixed objects. This ontology of the event, by nature impersonal, in my opinion structures the entire gravitational sphere of the theory of action and prevents an explicit, thematic treatment of the relation between action and agent, which the analysis nevertheless continually approaches.[38]

On Davidson's account 'anomalous monism' is the closest one can come to a theory of actions and events that respects the need for an adequate covering-law (or causal-explanatory) account while allowing for whatever is left over in the way of intentionalist, subjective, or agent-based modes of description. But it is also anomalous – as Ricoeur suggests – in the sense of preventing what its own analysis 'nevertheless continually approaches', that is to say, an understanding that would give due weight to the reasons and motives that agents (or speakers) typically adduce in order to explain their actions, words and beliefs.

Hence Davidson's frequent resort – along with others like Gilbert Ryle – to the device of recasting intentionalist predicates in a linguistic (usually adverbial) form which avoids such embarrassing commitments.[39] Thus 'by treating the intention as an adverb modifying the action, it is possible to subordinate it to the description of the action as a completed event'. Of course it may be argued, Ricoeur concedes,

> [that] actions are events, inasmuch as their description designates something that happens, as is suggested by the grammar of words; but no grammar allows us to decide between verbs that do not designate actions, such as 'err', and verbs that do designate actions, such as 'strike' or 'kill'. In this sense, the distinction between making happen and happening ... falls within the boundary of events. It is intention that constitutes the criterion distinguishing action from all other events.[40]

It seems to me that this also helps to account for the other great 'anomaly' in Davidson's work, namely the curious sense of non-alignment between his essays on the philosophy of action and events (where causal explanation plays such a large role) and his essays on language, truth, and semantics (where the main problem is precisely the lack of any adequate causal-explanatory approach). For in order to figure out the causes *and* the reasons why other people act, think and speak as they do we must possess at least some working grasp of the various motives, beliefs, and propositional attitudes that render their behaviour intelligible. This is what I referred to above as the second major 'gap' in Davidson's thinking: the absence of a theory that could plausibly explain how speakers and interpreters manage to converse on a basis of shared belief-attribution and linguistic or communicative uptake. That he purports to offer just such a theory – and thus to refute the more extravagant varieties of present-day sceptical-relativist doctrine – is in itself no reason for counting that attempt successful or (on his own Principle of Charity) taking the will for the deed. On the contrary: its failure is manifest in Davidson's retreat from the strong (though under-argued) claims put forward in 'The Very Idea of a Conceptual Scheme' to the minimalist semantics which is all that remains of that enterprise in 'A Nice Derangement of Epitaphs'. For at this stage – with the passage from 'prior' to 'passing' theories, and thence to his famous *tout court* pronouncement that 'there is no such thing as a language' – Davidson has abandoned every last resource for explaining how communicative uptake could possibly come about.

Here again there is something odd – not to say 'anomalous' – in Davidson's way of casually discarding his own (one would think more useful and productive) earlier insights. Thus in 'The Very Idea of a Conceptual Scheme' he argues convincingly that the sceptical-relativist doctrines of Quine, Whorf, Kuhn, Feyerabend *et al.* can appear plausible only if one takes it that semantics – rather than grammar, syntax, or logic – constitutes the bottom-line of linguistic understanding. That is to say, there exist certain well-known (and much-exploited) problems in explaining just how it comes about that speakers and interpreters manage to negotiate the often quite tricky differences of word-meaning across and between various idioms, registers, dialects, natural languages, and so on. But that they *do* so manage will seem much less of a mystery, Davidson suggests, if we switch attention to the various syntactic and logico-grammatical resources in the absence of which no language could achieve that level of functional or communicative adequacy which enables it to qualify as a language. Thus: '[w]hat forms the skeleton of what we call a language is the pattern of inference and structure created by the logical constants: the sentential connectives, quantifiers, and devices for cross-reference'.[41] In which case we can make a good start toward resolving the sceptical-relativist impasse through a simple recognition that, as Davidson puts it, 'syntax is so much more social than semantics'.

However, by the time of 'A Nice Derangement' he has backed off so far from this cardinal insight that it is now not just single words but uniquely occurring examples of the kind – *hapax legomena*, nonce-usages, malapropisms and so forth – which serve as a paradigm or test-case for linguistic understanding in general. For there is, according to Davidson, 'no word or construction that cannot be converted to a new use by an ingenious or ignorant speaker'. And if communicative uptake *just is* the ability to figure out what is meant by such token verbal occurrences – the ability, as Davidson says, 'to converge on a passing theory from time to time' – then clearly 'we have abandoned ... the ordinary notion of a language'.[42] Thus the way appears open to his further claim that 'in principle communication does not demand that two people speak the same language', since 'what must be shared is the interpreter's and the speaker's understanding of the speaker's words'.[43]

Nor is the issue much affected, Davidson thinks, if the speaker happens to hit on some word – whether through sheer ingenuity or mere ignorance – which corresponds to nothing in the interpreter's semantic or lexico-grammatical resources. For they can both get along just as well by trusting to that mixture of wit, luck, and wisdom which is all that 'knowing a language' comes down to on Davidson's minimalist account. Still this comes nowhere near explaining how language-users actually manage to communicate, even in those deviant cases (malapropism, aphasia, nonce-usages, idiolectal peculiarities) which might seem to bear out Davidson's point. For it is precisely when the interpreter's resources are stretched well beyond their customary scope – when she or he is hard put to discern any link between lexico-grammatical and utterer's meaning – that the whole complex business of 'knowing a language' is brought most crucially into play. In so far as he rejects this claim Davidson is effectively committed to the view that it *just doesn't matter* whether or not speakers make sense by any criterion of linguistic intelligibility available to us (or to them) at any level, no matter how basic, of shared communicative grasp.

Granted, there are passages in 'A Nice Derangement' which would seem, on the face of it, to support a quite different construal. Thus Davidson offers three propositions which he takes to characterize the standard (in his view mistaken) idea of what 'knowing a language' must entail. First: that any competent speaker or interpreter 'is able to interpret utterances, his own or those of others, on the basis of the semantic properties of the parts, or words, in the utterance, and the structure of the utterance'. Second: that in order to communicate successfully 'they must share a method of interpretation of the sort described in [item] 1'; and third: that 'the systematic knowledge or competence of the speaker or interpreter is learned in advance of occasions of interpretation and is conventional in character'.[44] Davidson thinks that (1) and (2) can be retained, albeit in modified form, but (3) has to go since it brings back a whole range of redundant notions – 'systematic knowledge', 'prior theories', 'learning in advance', speaker-'competence', linguistic 'conventions', and so forth – which are simply not required on the minimalist account.

However, it is far from clear that Davidson can dump item (3) altogether while hanging onto what he needs of items (1) and (2) in order to make his theory in some degree plausible. That is to say, he is still committed – in principle at least – to some version of the Tarskian truth-based compositional semantics that would seek to interpret each and every meaningful utterance of a language in terms of its 'structure' as recursively defined through a process of analysis into 'properties', 'parts', and 'words'. Such is the 'method of interpretation' which Davidson thinks quite sufficient for the purpose without dragging in all that otiose talk of 'systematic knowledge', 'conventions' and so forth. But the problem remains as to what could possibly constitute such a 'method' – or such a basis of shared speaker- and interpreter-'competence' as described (semi-approvingly) in (1) above – if it did not involve some element of 'learning in advance', or some more or less 'systematic' means of interpreting familiar or novel modes of utterance. Thus one may suspect that Davidson is fudging the issue when he writes that the first two principles 'survive when understood in rather unusual ways', but that the third 'cannot stand' although it is unclear just 'what can take its place'. For in fact (1) and (2) are pretty much synonymous with (3), or at any rate with that portion of (3) that simply explicates 'competence' – or 'successful communication' – in terms of 'systematic knowledge' and 'learning in advance'. Where the argument gets skewed, I would suggest, is when Davidson then goes on to imply that such knowledge can only be 'conventional' in character, thus landing his opponent with all the well-known difficulties canvassed in 'The Very Idea of a Conceptual Scheme'.

However there is no reason to accept this smuggled-in premise that any appeal to prior theories, 'systematic knowledge' or whatever, will necessarily be just a matter of convention and hence open to the sorts of knock-down anti-conventionalist argument that Davidson standardly deploys. On the contrary, as he remarks in an earlier essay: 'convention does not help to explain what is basic to linguistic communication, though it may describe a usual, though

contingent, feature'. And again: 'philosophers who make convention a necessary element in language have the matter backwards.
The truth is rather that language is a condition for having conventions'.[45] (To which one might add, with good Davidsonian warrant,
that attributing truth-values to a speaker's utterances is in turn the
condition for having both a language and those language-dependent conventions.) But in that case it is hard to see how Davidson can
be justified – even on the widest, most generous principle of charity
– when he takes it as read, in 'A Nice Derangement', that talk of
prior theories or 'systematic knowledge' necessarily involves some
conventionalist idea of linguistic or interpretive competence. In fact
it looks very much as if Davidson is attempting to skewer his
notional opponents on the horns of a false dilemma.

EMPSON AND DAVIDSON ON 'KNOWING A LANGUAGE'

Empson, I think, points the best way beyond this apparent dead-
end in philosophical semantics. Where he succeeds – and where
Davidson ultimately fails – is in developing a theory which on the
one hand incorporates a generalized 'principle of charity' (i.e., the
Davidsonian default principle that speakers will *likelier than not*
make sense in rationally explicable ways), while on the other hand
acknowledging the sheer range and complexity of what actually
goes on in the process of communicative uptake. For the trouble
with Davidson's theory, as we have seen, is that it leaves no room
for the crucial distinction between non-standard (for instance metaphorical, ironic or otherwise deviant) forms of meaningful utterance and cases where the speaker has gone completely off the rails
and we are left to figure out his or her intentions in the absence of
any plausible semantic or lexico-grammatical account. That is to
say, it goes so far toward maximizing truth-values – or giving
speakers the benefit of the doubt – that at the limit (as for instance
with the more extreme cases of malapropism or idiosyncratic
nonce-usage) one is not so much 'interpreting' their words as simply ignoring the gist of what they say in favour of some altogether
different, better or more 'charitable' substitute meaning.

   Thus one construal of Davidson's famous dictum 'there is no such
thing as a language' is that it treats such instances of scratch interpretation – involving a mixture of inspired guesswork and contextual (extra-linguistic) clues and cues – as far more typical of human

interactive behaviour than anything that linguists (or philosophers of language) have so far managed to come up with. From which it follows that there is no difference in principle – no linguistically or philosophically interesting difference – between cases of metaphor, linguistic inventiveness, irony or innovative usage and cases of downright nonsensical expression where speaker's meaning (or utterer's intent) must be taken as *completely unrelated* to anything that the interpreter may bring to it in the way of 'prior theories' or shared linguistic competence. In fact this is the only way that Davidson's theory can itself be made to yield any kind of intelligible sense. For there is otherwise no rational construal of his claim that speakers can 'mean' whatever they like by the use of some word or expression – can give it what he calls, oxymoronically, a 'novel literal meaning' – just so long as interpreters enjoy the same licence of *ad hoc* intuitive guesswork.

Even by applying this Principle of Charity in accordance with Davidson's prescription one still comes up with a 'theory' of language and interpretation that fails to meet the most basic requirements of rational intelligibility. What it lacks, in short, is any means of explaining (1) how words can take on different senses (novel, unfamiliar, metaphorical, idiosyncratic and so forth) in certain likewise unusual contexts of usage; (2) how we can none the less interpret such words with a fair degree of assurance by analogy with other (more commonplace) modes of linguistic or communicative grasp; and (3) how these instances differ from cases of, for example, full-blown malapropism or chronic aphasia precisely in so far as they can still be *interpreted*, rather than stretching the Principle of Charity to a point where *every* sort of deviant utterance requires such treatment, aphasic ramblings or meaningless word-strings included.

One can therefore understand why critics like Papineau see something rather less than charitable in Davidson's theory. For on this account it is not just theorists and philosophers who find themselves out of a job, having patiently laboured at their grandiose conceptions and now – thanks to Davidson – acknowledged the pointlessness of all such endeavours. More than that, it is the case for all language-users (Davidsonian 'speakers' and 'interpreters') that they can perfectly well get by on just the sort of *ad hoc* or minimalist basis – joined to an all-encompassing Principle of Charity – that applies (or so he argues) in the case of Sheridan's Mrs Malaprop. For if this is all there is to 'knowing a language' then a good many linguists, philosophers, and cognitive psychologists (Chomsky among them) have clearly been over-impressed by that particular capacity. Still less could they be justified in mounting large claims about human intelligence, ethical values, political responsibility and so forth, on so desperately thin and exiguous a basis as that of mere language-acquisition or our everyday 'competence' in figuring out the gist of other people's speech-acts. For as Davidson treats it, this process is pretty much the same whether we are stretching our minds around some complex, novel, provocative, or humanly challenging utterance or otherwise just doing the best we can for some transient (perhaps pathologically induced) instance of deviant usage.

This is why Papineau advances his 'Principle of Humanity' as a better – indeed a more charitable – option than Davidson's Principle of Charity.[46] What his approach leaves open (and what Davidson's theory closes off) is the standing possibility that other people may *sometimes* be wrong in their beliefs; that we may ourselves at times have reason to count them mistaken while nevertheless correctly interpreting their words; and that on other occasions we may be driven to give up on this optimizing strategy and count not only their beliefs but also their utterances incapable of any adequate (sense-preserving) interpretation. In the second case (that of false beliefs), the most humane *and* charitable course may be to offer a causal-explanatory account of why they should say such things – and hold such beliefs – without, as Davidson puts it, falling into the kind of 'massive error' that would place them simply beyond reach of rational understanding. Thus they might be in the grip of some powerful prejudice or some socially sanctioned (for example, religious) belief-system that made it well-nigh impossible for them to break through to a more adequate, less partial or distorted view of things. In other words the Principle of Humanity, as Papineau construes it, enables us to maximize the *rational* content of their meanings, intentions or beliefs while regarding the truth-content of their utterances as still open to criticism or outright rejection.

This in turn makes it possible to distinguish the second from the third type of case as described above. For with the latter – where properly linguistic understanding just cannot get a hold – the interpreter has a choice between two strategies. One, following Davidson, is to adopt a wholesale Principle of Charity requiring that here also (as in cases of false but intelligibly expressed and understandably arrived-at belief), we should seek to optimize truth-content by imputing some intention – or utterer's meaning – which at the limit may entail a complete disregard for the lexical, grammatical, or logico-semantic resources of our or their language. This amounts to a vote of no confidence in 'language' – as standardly conceived – and a recommendation that we go all the way with a minimalist-semantic line according to which speakers always (necessarily) say what they mean and mean what they say since we could never be justified – on charitable grounds – in supposing otherwise. The second strategy rejects this approach as involving a reductive (scarcely even 'charitable') idea of the complex relationship that exists between truth, belief, utterer's meaning, communicative import, and what properly counts as 'knowing a language'. It holds that we can do no favour to ourselves (or others) if we take that Davidsonian line of least resistance which counsels us to maximize the intelligibility (or truth-content) of utterer's meaning, if need be – where the purport is excessively obscure – by pretty much ignoring the sense of their words as given by a prior (hopefully shared) theory of interpretation. On this account, 'charity' will tend to work out as a distinctly one-sided affair, a gambit that allows all manner of inventive (belief-optimizing) freedoms on the resourceful interpreter's part while treating the speaker – or 'native informant' – as one whose utterances, meanings, and beliefs can always be subject to a kind of negotiated trade-off on terms that make sense by the interpreter's lights. And as Davidson sees it there is no difference in principle between the sorts of allowance that we learn to make in cases of chronically deviant usage or skewed communication (as with Mrs Malaprop) and those far more commonplace instances – metaphors, novel turns of phrase, idiomatic expressions – where 'prior theories' are supposedly of no use at all, and where even 'passing theories' possess no more than a localized, potshot utility.

Thus the Principle of Charity ironically ends up by suggesting that both parties – speaker and interpreter alike – have nothing to rely on save 'wit, luck, and wisdom' plus, on the good-willed interpreter's side, a large (indiscriminate) measure of tolerance for deviant or nonsensical modes of utterance. And indeed it is hard to see, on Davidson's theory, how any viable distinction could be drawn between those manifold types of deviant expression (from downright nonsense, *via* false or irrational beliefs, to localized errors of usage on the one hand and metaphors, innovative speech-styles, and idiolectical variations on the other) which require something more than an across-the-board strategy for optimizing utterer's intent. This is why Papineau proposes 'humanity', rather than 'charity', as the best working principle for a theory of interpretation that would accord a genuine (and not just a notional) respect to the meaning, rationality and truth-content of various sample-utterances. For it is not, after all, the most generous of attitudes that would count other people 'necessarily' (i.e., for all we can know) 'right in most matters' if this involves simply levelling the difference between truth and falsehood, sense and nonsense, or linguistic errors and linguistic innovations. What is required of an adequate theory – as also of any competent speaker/interpreter – is the ability to distinguish between these cases and, where appropriate, assign some humanly intelligible reason (maybe of a causal-explanatory nature) for just why the speaker should talk and think that way. This approach is *both* more humane and more charitable since it treats speakers and interpreters alike as doing their best – sometimes against considerable odds – to achieve better understanding on the basis of a shared truth-seeking interest that doesn't rest content with a merely formal (Davidsonian) concordat or agreement not to differ.

Still there is a need to explain what the Principle of Humanity amounts to in detailed (theoretical and applied) terms if it is to offer something more than Davidson's kind of all-purpose optimizing strategy. I have suggested already – but the point will bear repeating – that no philosopher has done as much as William Empson to show how the principle might actually work out as a matter of interpretative method and practice. Some of his examples touch directly on issues in the philosophy of mind, knowledge, and language. One is the keyword 'sense' as deployed – often to remarkably subtle effect – by writers such as Shakespeare, Pope, Jane Austen, and Wordsworth.[47] Its implications range all the way from the rock-bottom empiricist (or phenomenalist) doctrine 'sense = only what comes to us in or through the senses', *via* Pope's more 'sensible' or moderate variations on the theme ('the good humour of a reasonable man' . . . 'that "common sense" which is to become adequate to the task of criticism'), to Wordsworth's mystical-pantheist 'language of the sense', that is, his deployment of 'Type IV' equations to suggest a transcendent, visionary state of mind beyond all the vexing antinomies of plain-prose (commonsense) reason. Empson is worried about the Wordsworth case where, as he puts it, 'the whole poetical and philosophical effect comes from a violent junction of sensedata to the divine imagination given by love, and the middle term is cut out' (*Complex Words*, p. 296). Thus the reader may well feel tempted to complain that 'what is jumped over is "good sense"; when Wordsworth has got his singing robes on he will not allow any mediating process to have occurred' (p. 304).

In fact Empson had long been puzzled by this aspect of Wordsworth's language, beginning with some passages of hard-pressed (even 'niggling') analysis in *Seven Types of Ambiguity* where he tried to figure out how the poetry could achieve such a powerful emotive effect while not making sense – or arguing its case – in coherent grammatical or logico-semantic terms.[48] Most likely it was this instance, along with some other (less exalted) examples of the kind, that led Empson to abandon his vaguely inclusive talk of poetic 'ambiguity' and adopt the more exacting, philosophically accountable mode of analysis developed in *Complex Words*. Thus '[i]t does not seem unfair to say that he [Wordsworth] induced people to believe he had expounded a consistent philosophy through the firmness and assurance with which he used equations of Type IV; equations whose claim was false, because they did not really erect a third concept as they pretended to'. Nevertheless, 'in saying this I do not mean to deny that the result makes very good poetry, and probably suggests important truths' (p. 305).

This should not be seen as just a shuffling or evasive compromise 'solution' on Empson's part. Rather, it provides a good working instance of that 'Principle of Humanity' that enjoins us to optimize truth-content or rational sense so far as possible, and only then to cast around for some other, for example, 'emotive' explanation of why speakers (poets included) and interpreters (readers of poetry among them) may sometimes get along without. It is worth contrasting this with the kind of interpretation that might result from applying the Principle of Charity in its wholesale Davidsonian form. What would then drop out altogether is that sense of inherent strain or resistance – that awareness of language being pushed up against the limits of paradoxical assertion – which Empson finds 'philosophically' a problem with Wordsworth but also a chief source of his 'poetical' power. This is to not say – far from it – that the only fit state of mind for a reader of poetry is one that suspends all the normative requirements of logic, consistency and truth and which accepts any kind of 'deep' (paradoxical) wisdom just so long as it carries a sufficient affective charge. Indeed it is Empson's main purpose in *Complex Words* to challenge this doctrine in its various philosophical and literary-critical forms. These ranged from the emotivist theory adopted by I.A. Richards (and by ethical theorists like C.L. Stevenson) as a hedge against logical positivism to the rhetoric of 'paradox', 'ambiguity, 'plurisignification', and so on, by which the New Critics sought to erect a *cordon sanitaire* around poetry's privileged domain.[49] In *Seven Types* Empson took a dissident stand by pushing his method of verbal analysis as far as it would go, and also – contrary to orthodox precept – by paraphrasing freely in the effort to tease out poetic ambiguities. All the same he came to feel that the method itself was unfortunately prone to encourage just the kind of irrational mystery-mongering that Empson so disliked about the New Criticism and its various latter-day offshoots. Hence, as we have seen, his shift of analytical approach from Ambiguity to Complex Words, that is to say, from a somewhat catch-all idea of multiple meaning in poetry to a theory which fitted those meanings into 'definite structures' of logico-semantic entailment. For it is the virtue of this approach, Empson thinks, that it offers some hold for rational understanding even in cases – like Wordsworth's 'Type IV' equations – where the doctrine concerned is obscure to the point of very nearly resisting such analysis.

On the alternative (e.g., New Critical) account these problems drop out of view since paradox is taken as a veritable touchstone of poetic value, and also – by the same token – as a sign that these are regions where the rational prose intellect should fear to tread. (Thus Cleanth Brooks: 'what Wordsworth had to say demanded the use of paradox . . . could not be said effectively without paradox').[50] So it was that the more doctrinally minded of the New Critics – Brooks and W.K. Wimsatt among them – saw fit to attack the 'heresy' (or 'fallacy') of paraphrase as one of those bad habits by which interpreters avoided a close engagement with the sacrosanct 'words on the page'.[51] Indeed they took issue with Empson on just these grounds, despite acknowledging the brilliance of his work and its formative influence on their own ways of reading. What they found most vexing about *Seven Types* was its outlook of sturdy commonsense rationalism joined with its resistance to any sort of doctrine – aesthetic, ethical, or religious – which would seek to cut poetry off from from our 'normal waking habits' (Empson's phrase) as speakers and interpreters of language.[52]

Empson had exactly the opposite reason for suspecting that the book had exerted a harmful influence, not least – despite their express misgivings – on the New Critics themselves. For it could hardly be denied that Empson's most powerful and striking examples were those grouped together under the seventh Type, that is to say, cases of 'deep-laid psychological conflict' where the two or more senses of some word, phrase or passage were felt to embody an extreme clash of attitudes which most often found expression in sharply paradoxical form. What made matters worse, from Empson's later viewpoint, was that these conflicts were often presented as stemming from the poet's struggle with the harsher implications of orthodox Christian belief, as for instance in his reading of Hopkins's 'The Windhover' and his extraordinary line-by-line heterodox commentary on Herbert's 'The Sacrifice'.[53] Empson's point had been to interpret these poets as engaged – at whatever 'unconscious' level – in a desperate attempt to make humanly intelligible sense of a religion which required its adherents to accept such mind-wrenching paradoxes as the doctrine of vicarious Atonement, or God's 'satisfaction' through the suffering and death of Jesus Christ.

Thus the power of these poems came of their dealing with certain 'complicated and deep-rooted' notions which might in the end defeat the best efforts of logico-semantic analysis. Nevertheless that power could only be felt if the reader (like the poet) managed to keep some rational 'machinery' in play, some background of normative judgement against which to interpret the various contradictions involved. Simply to accept them as paradoxes – as profound truths beyond reach of plain-prose reason – was, according to Empson, not a fit state of mind in which to appreciate the poetry. Indeed it might be seen – though of course he doesn't make the point in these terms – as just another, more specialized literary variant of the Davidsonian Principle of Charity. That is, it requires that we suspend all those normally operative processes of thought whereby – on any adequate account of linguistic competence – we are enabled to distinguish the various orders of literal, metaphoric, paradoxical and (at the limit) nonsensical or meaningless utterance.

In *Seven Types* Empson devotes some rather fretful pages to the question how far analysis should go in the attempt to make rational sense of otherwise obscure (but often powerful or highly charged) modes of expression. His own belief – while offered as perhaps just 'a useful prejudice with which to approach the subject' – is that 'although such words appeal to the fundamental habits of the human mind, and are fruitful of irrationality, they are to be

èxpected from a rather sophisticated state of language and of feeling'.[54] A couple of pages on he makes the same point by asserting,
more simply, that 'any contradiction is likely to have some sensible
interpretations', and that 'if you think of interpretations that are not
sensible, then it puts the blame on you'.[55] But there is still a problem
as to how any line can be drawn (and with what degree of assurance) between 'contradictions' that are felt to require some further,
hopefully clarifying effort of logico-semantic analysis and 'paradoxes' where the meaning is assumed to lie too deep – or too far
back – for any such method to really get a hold. And this question
became more urgent for Empson as he witnessed the rise of a
powerful academic orthodoxy – that of the New Criticism – which
took his own work (or a selective reading of passages from *Seven
Types*) as a source for the doctrine that poetry *just was* 'paradoxical'
through and through.[56] From here it was no great distance to those
modes of geared-up verbal exegesis, often with a thinly disguised
theological subtext, which located the chief value of poetry in its
resistance to any kind of rational understanding arrived at through
logical analysis or plain-prose paraphrase.

His patience very often reached breaking-point when reviewing
books by Wimsatt, Brooks and others where this method was put
into practice, usually (he thought) with disastrous effects upon
their habits of literary and moral judgement. Thus in 1963, with a
rueful backward glance to his pages on Herbert's 'The Sacrifice': 'it
strikes me now that my attitude was what I have come to call
"neo-Christian"; happy to find such an extravagant specimen, I
slapped the author on the back and urged him to be even more
nasty'.[57] And again, of Brooks's essays in *The Well Wrought Urn*: 'he
is too content to find the intellectual machinery of a fine and full
statement in the poem; there is enough irony and paradox and so
on, he feels, for the meaning to be made profound; this is true, but
you still need to ask whether the machine worked the right way'.[58]
Worst of all – and his patience completely ran out at this point –
was Wimsatt's anti-intentionalist argument that a close enough
attention to the poem as 'verbal icon' (to its inwrought structures of
ambiguity, irony or paradox) could best be relied upon to stop
readers from raising such awkward and, to his mind, illegitimate
questions of authorial intent. 'Consider the Law', Empson remarks,
'which . . . recognizes amply that one can tell a man's intention, and
ought to judge him by it. Only in the criticism of imaginative
literature, a thing delicately concerned with human intimacy, are

we told that we must give up all idea of knowing his intention'.[59] The main use for the Wimsatt Doctrine was (he thought) to smuggle in pious revisionist meanings – most often of a 'deep' paradoxical character – which could then be passed off as truths beyond reach of the meddlesome prose intellect. And this method clearly had its pedagogic uses in the USA at a time (the mid-1950s) of intensifying pressure to conform with the dictates of a right-wing social, political, and religious agenda. Thus in Wimsatt's case 'the idea that a piece of writing which excited moral resistance might be a discovery in morals, a means of learning what was wrong with the existing system, somehow cannot enter his mind'.[60] And yet, as Empson remarks in a tone of justified exasperation: 'surely this has often happened and provides the only interesting question for his article'.

I think that this helps to explain why Empson should have puzzled so long and hard in *Seven Types* about the question whether cases of poetic paradox might stand beyond reach of logical analysis, or again – conversely – how far they could be viewed as 'contradictions' which actually required and justified such treatment. In fact he has some dense but rewarding paragraphs on the topic which clearly look forward to his later, more elaborate theory as developed in *The Structure of Complex Words*. I shall quote from them at length since they offer – to my mind – a valuable gloss both on Empson's thinking and on the relevance of his work to issues in current analytic (and 'post-analytic') philosophy. 'When a contradiction is stated with an air of conviction', he writes,

> it may be meant to be resolved in either of two ways, corresponding to thought and feeling, corresponding to knowing and not knowing one's way about the matter in hand. Grammatical machinery may be assumed which would make the contradiction into two statements; thus 'p and –p' may mean: If $a=a_1$, then p; if $a=a_2$, then –p'. If a_1 and a_2 are very different from one another, so that the two statements are fitted together with ingenuity, then I should put the statement into an earlier type; if a_1 and a_2 are very like one another, so that the contradiction expresses both the need for and the difficulty of separating them, then I should regard the statement as an ambiguity of the seventh type corresponding to thought and knowing one's way about the matter in hand. But such contradictions are often used, as it were, by analogy from this, when the speaker does not know what a_1 and a_2 are; he

satisfies two opposite impulses and, as a sort of apology, admits
that they contradict, but claims that they are like the soluble
contradictions, and can safely be indulged . . . . One might think
that contradictions of this second sort must always be foolish, and
even if they say anything to one who understands them can quite
as justifiably say the opposite to one who does not. But, indeed,
human life is so much a matter of juggling with contradictory
impulses (Christian-worldly, sociable-independent, and suchlike)
that one is accustomed to thinking people are probably sensible if
they follow first one, then the other, of two such courses; any
inconsistency that it seems possible to act upon shows that they
are in possession of the right number of principles, and have a fair
claim to humanity.[61]

Compare this with the prospects afforded by a Davidsonian (minimalist-semantic) theory and you will have some idea of that theory's shortcomings as a purported account of how speakers and
interpreters remarkably (but regularly) manage to do what they do.
Indeed Empson's approach here – as throughout *Seven Types* – is a
good deal closer in method and ethos to the 'Principle of Humanity' that Papineau proposes as a substitute for Davidson's all-
purpose Principle of Charity. That is, it takes account of those
various complicating factors – social, cultural, attitudinal, disposi-
tional, depth-psychological – which may require speaker and interpreter alike to stretch their minds around problems of
communicative grasp undreamt-of in Davidson's simplified conception.

Thus it is more than just a chance coincidence of phrase when
Empson remarks that this capacity for 'juggling with contradictory
impulses' may be taken as a sign that speakers are 'in possesssion
of the right number of principles' and have 'a fair claim to humanity'. What links his approach with Papineau's – despite their
obvious differences of interest – is the idea that an adequate understanding of other people's motives, meanings, and intentions may
involve something more – though also (to be sure) nothing less –
than the effort to make best sense of their words on a truth-optimiz-
ing principle. That is to say, we need to start out by assuming that
they (like us) are possessed of sufficient logico-semantic 'machinery' for it to be most likely the case that their utterances will
bear some 'sensible' interpretation, some meaning that doesn't on
the face of it reduce to downright contradiction, illogicality, or

other such blocks to communicative uptake. Moreover there is plenty of evidence that speakers and interpreters are better equipped in this regard than Davidson is able to explain through his blanket appeal to a Principle of Charity which counts them necessarily 'right in most matters' without much attempt to show in any detail how this optimizing strategy works.

3

Doubting Castle or the Slough of Despond: Davidson and Schiffer on the Limits of Analysis

MINIMALIST SEMANTICS AND THE 'NO-THEORY THEORY OF MEANING'

At present there would seem to be two main camps in Anglo-American philosophy of language, the split falling out much as Richard Rorty described it in the Preface to his 1967 anthology *The Linguistic Turn*.[1] His editorial policy there was to give an even-handed coverage to both sides of the emergent dispute while suggesting that their differences could not be resolved and therefore that the only way forward was to adopt a sensibly pragmatist view which entitled one to pick and choose without any need to take sides. On the one hand were those 'analytical' types in the Frege–Russell line of descent who took it that 'ordinary language' was too fuzzy, imprecise, or ambiguous to provide an adequate basis for the conduct of philosophical enquiry. It could be rendered fit for that purpose only through a rigorous analysis of its underlying logical grammar, or a method – such as Russell's Theory of Descriptions or Frege's canonical distinction between Sense and Reference – for effectively dispelling the manifold sources of 'commonsense' error and illusion.[2] On the other side were those in the 'ordinary language' camp – influenced chiefly by Wittgenstein and Austin – who rejected the idea that language could or should be subject to such forms of abstract logical regimentation. In their view, as Austin famously expressed it, our 'common stock of words' embodied all the distinctions, nuances, connections and refinements that speakers had 'found worth marking in the lifetimes of many generations'. From which it followed that 'these surely are likely to be more numerous, more sound, . . . and more subtle, at least in all ordinary and reasonably practical matters, than any that you and I are likely to think up in our arm-chairs of an afternoon – the most favoured alternative method'.[3]

To Rorty this seemed just one more example of the kinds of dilemma that philosophers typically got into by supposing that there must be a right way of doing things and that theirs was the 'method' (or, in Austin's case, the modestly unmethodical approach) by which best to do it. His own work up to this point had been largely 'analytical' in character, or addressed to problems within and around that first (Frege–Russell) line of descent. But thereafter – that is to say, in his writings subsequent to *The Linguistic Turn* – he swung right across to a pragmatist view which left little room for such specialized concerns. Thus Rorty now argued that philosophy is not a 'constructive' or problem-solving exercise; that the analytic enterprise had reached a dead-end with the difficulties uncovered by 'post-analytical' thinkers like Quine, Sellars, and Goodman; and hence that the most useful ('edifying') job of work for philosophers was to help this beneficial process along by debunking the discipline's old pretensions and maybe – once in a while – coming up with some novel metaphor or narrative slant on its own history to date.[4]

Another route 'beyond' analytic philosophy is that taken by Donald Davidson in a series of influential essays, among them 'On the Very Idea of a Conceptual Scheme' and (more recently) 'A Nice Derangement of Epitaphs'.[5] Davidson has shifted ground to some extent during the roughly ten-year period that separates these two publications. Nevertheless, one can see how he travelled the path from a truth-based (Tarskian) compositional semantics to a position that Rorty can cheerfully endorse – if somewhat to Davidson's discomfort – as one more feather in the gathering wind of post-analytical fashion.[6] For as Davidson now sees it the only 'theory' that is needed is one that effectively puts itself out of business by taking each utterance as it comes, attributing intentions on a one-off (*ad hoc* or intuitive) basis, and assuming that context – or circumstantial cues and clues – can make up any deficit supposedly created when we drop all that otiose philosophic talk about 'knowing', 'possessing', or 'sharing' a language.[7] At which point the question arises: why adopt this line of last resort when there exist alternative approaches with a far greater claim to technical refinement and conceptual-explanatory grasp? But of course it is Davidson's belief – and his main reason for taking this minimalist tack –

that those approaches have not *and in principle cannot* come up with anything remotely resembling an adequate theory of language and interpretation. And this because all the principal contenders (that is to say, those that fall within Davidson's purview) fail to meet his first desideratum, namely the requirement of explaining how it is that we interpret novel – hitherto unmet-with – items of speech behaviour. Rather they are subject to what Davidson sees as an inverse law of diminishing returns whereby the most (apparently) powerful or elaborate theories are of least use in coping with just such cases.

However, it is also worth noting how the target has shifted in the course of Davidson's visions and revisions from 'On the Very Idea of a Conceptual Scheme' to 'A Nice Derangement of Epitaphs'. In the earlier essay – and in other pieces written at about that time – he maintained some resistance to those forms of holistic or contextualist theory which relativized meaning and truth to the entire set of sentences, propositions or beliefs accepted as meaningful or true at some given time by some given community of speakers.[8] This was because they ruled out the kind of recursive compositional analysis that would assign truth-conditions to *particular sentences* in context, and thus provide at any rate the formal basis for a generalized interpretative theory.[9] Stephen Schiffer offers the following brief account of what such a theory might look like in his book *Remnants of Meaning*. (I take the description from Schiffer, and not more directly from Davidson, for reasons that will soon become apparent.) Thus:

> a *compositional (truth-theoretic) semantics* for a language L is a finitely statable theory that ascribes properties to, and defines recursive conditions on, the finitely many vocabulary items in *L* in such a way that for each of the infinitely many sentences of *L* that can (in principle) be used to make truth-evaluable utterances, there is some condition (or set of conditions) such that the theory entails that an utterance of that sentence is true iff that condition (or a certain member of the set) obtains.[10]

Moreover, it is assumed that some such theory is both implicit in our normal (everyday) competence as language-users and prerequisite for any philosophical approach that would explain that competence. In Schiffer's words: '[i]t would not be possible to account for a human's ability to understand utterances of indefinitely many novel sentences of a language without the assumption that that language had a compositional semantics'.[11]

These statements correspond closely to Davidson's earlier position, set out in essays like 'The Very Idea', 'Truth and Meaning', and 'The Method of Truth in Metaphysics'.[12] What is notable about them in the present context of argument is just how far he had shifted away from that position by the time he came to write 'A Nice Derangement of Epitaphs'. For it is now Davidson's firmly held belief that no such theory could ever provide a useful starting-point – let alone a 'method' – for interpreting novel utterances. Indeed it strikes him as merely self-evident that their 'novelty' is enough to place them *by very definition* beyond reach of any generalized (truth-theoretical) account. On this view, what constitutes a novel speech-occurrence is the fact of our never before having met with some particular expression or form of words in some particular (uniquely specified) context of usage. Since there is, after all, 'no such thing as a language' – nothing that could answer to his earlier set of conditions for knowing, possessing, interpreting, or sharing such a thing – Davidson concludes that meanings and contexts are likewise as many and various as the speech-situations in which we find ourselves from one moment to the next. One may suspect that the Tarskian (truth-theoretical) approach remains always somewhere in the background, to be drawn upon tacitly where needed in order to reduce the otherwise extreme credibility-gap created by this minimalist account. But there is still the question as to why Davidson should have felt himself driven to adopt a line of strategy – a 'transcendental argument to end all transcendental arguments', as Rorty puts it – which seems so much at odds with his own earlier position.[13] One answer, I would suggest, was his failure to extract any more substantive (useful or applicable) theory of truth and meaning from the formalized Tarskian account.[14] Hence his resort to the opposite extreme, that is, to a notion of utterer's meaning that would somehow ensure communicative uptake quite aside from any knowledge – on the speaker's or the interpreter's part – of what constitutes 'a language' and its relevant structures of lexico-grammatical or logico-semantic grasp.

To this extent, Davidson's dilemma can be seen as typifying a widespread movement of retreat among philosophers in the post- (or ex-) analytical tradition. Hilary Putnam offers perhaps the most striking instance, since in his case the stages are clearly marked out through a series of books over the past two decades that chronicle his sense of progressive disenchantment with most aspects of the old enterprise, and his coming around to a compromise position – so-called 'internal realism' – which amounts to a fairly minor variant on the pragmatist or holist theme.[15] If 'realism' remains a part of this programme then it is only in order to mark his distance from those other, less 'robust' versions which – like Rorty's – lay themselves open to the standard forms of knockdown anti-relativist argument. Thus Putnam goes so far as to claim that his is basically a Kantian intervention for our times, seeking to turn philosophy away from its deluded ('metaphysical') realist beliefs and recall it to a decently scaled-down sense of its own proper scope and limits. But this claim will scarcely seem convincing if one considers the extent of Putnam's visions and revisions since his work of the mid-1970s. And it is no more convincing, in Davidson's case, when the argument rests on a supposed return to the context of everyday pragmatic speech- behaviour as contrasted with the kinds of high-level formalized (or meta-linguistic) concern which marked that earlier phase of enquiry. For the appeal to commonsense carries little weight if it means giving up any serious attempt to provide an adequate theoretical account of what speakers and interpreters manage to achieve in everyday communicative contexts.

I suggested above, when quoting a passage from Stephen Schiffer's *Remnants of Meaning*, that his book was of particular interest with respect to these issues in present-day philosophy of language. For that passage described what Schiffer had once taken as a paradigm of philosophic method, but had now come to regard – some twenty years on – as a failed and bootless endeavour.[16] Jerry Fodor captured the tone well enough in his review of the book: 'you can read *Remnants of Meaning* as a philosopher's $8\frac{1}{2}$, an analytical *Baby, It's Cold Outside*; imagine *Either/Or* rewritten by Tarski and you'll have the feel of it'.[17] Schiffer himself is in some doubt as to whether his revised view of things could aptly be described as 'defeatist' or 'despairing'. The former adjective he thinks somehow inappropriate, although there is now – on his own submission – no hope of carrying through that original programme despite all the objections that have been levelled against it. As for *despairing*, 'that is a more difficult question, and one that I care very much about' (*Remnants*, p. 271). Perhaps (he suggests in the book's closing paragraph) there are new questions that might yet be raised, and even the possibility of a constructive answer 'in some alliance with cognitive science'. But one can sympathize with Fodor – a leading proponent of the cognitivist view – in finding this conclusion oddly out of joint with the rest of Schiffer's arguments.[18] For these are devoted in very large part to a sceptical assault upon just about everything that would count in favour of a cognitivist approach to issues of language, meaning, and interpretation.

In Schiffer's end is also his beginning, since the book starts out by listing those items of analytic faith which he now feels compelled to renounce. Together they make up the theory of Intention-Based Semantics (IBS for short) whose salient features are those set forth – with a view to impending demolition – in the passage from Schiffer quoted above. These include: (1) the postulated existence of 'semantic facts' that relate marks or sounds to determinate properties of meaning, truth, and reference; (2) the claim that each natural language has a compositional meaning-theory, a theory that is 'finitely statable' and can be applied recursively to specify a meaning for every word, sentence, or syntactic construction of the language; (3) a truth-theoretic (i.e. Tarskian) semantics which assigns a relevant truth-condition (hence a meaning) for each such utterance in context; and (4) the argument that there can be no accounting for our everyday linguistic competence – for '[our] ability to understand utterances of indefinitely many novel sentences of a language' – except on the basis of a compositional semantics as specified in items (1) to (3).

So much for the first-order components of Schiffer's (now abandoned) theory, those having to do with propositional content or the meaning of sentences that involve no element of speaker-related attitude or belief, and which can thus be analysed – along Tarskian lines – in a purely extensionalist mode. What remains to be accounted for is the relation between these primary components of sentence-meaning and those other, more complex forms of utterance whose meaning is rendered opaque (or resistant to any such extensionalist account) through their entailing some strictly ineliminable appeal to mind-states, beliefs, assenting or dissenting dispositions, and so forth. It is at this point that Schiffer introduces a further series of hypotheses which he once (happy time!) thought sufficient to deal with the problem but now finds woefully inadequate. Chief among them is (5) the 'relational theory of propositional attitudes' which treats sentences as objects of belief and beliefs as objects of a semantic theory whereby their intentional (speaker-related) meaning can itself be analysed in truth-conditional terms. Thus: 'believing is a relation to things believed, to values of the variable '*y*' in the schema '*x* believes *y*', which things have features that determine the intentional features of beliefs' (*Remnants*, p. xvi). On this account there is no reason to suppose – like (at any rate the early) Quine – that extensionalism demands the complete elimination of all such 'mentalist' residues, that is to say, all talk of meanings or beliefs which cannot be directly cashed out in extensional terms.[19] What is required, rather, is a treatment of propositional attitudes that renders them perspicuous (or non-opaque) by bringing them within the extended range of a compositional semantics. For if such a theory holds good for any language, then it follows – in Schiffer's words – that ' "believes" must be treated in that semantics as a semantic primitive', and that arguably 'the only tenable way this can be done is to treat "believes" as a relational predicate true of a believer and what he believes' (*Remnants*, p. xvi).

As I have said, the remainder of Schiffer's book is a kind of reverse *Pilgrim's Progress*, a loss-of-faith narrative whose endpoint is either Doubting Castle or the Slough of Despond. In this respect it serves as an unusually forthright statement of difficulties that others – Davidson among them – have likewise encountered in attempting to derive a workable theory of natural-language interpretation from the Tarskian formalized or truth-theoretic approach. With Davidson, the story appears to possess a more upbeat, or at any rate less bleak and tormented outcome. However one may doubt that the prospects are really much brighter for Davidson's proposed solution, his suggestion that we should travel light – philosophically speaking – and just get along without the dubious benefit of 'prior theories', 'knowing a language' and so forth.[20] The source of its appeal, for him as for some commentators, is that it seems to put the emphasis squarely back on the context of natural-language communicative utterance, that is to say, on the ways in which ordinary speakers and interpreters – as distinct from theorists or philosophers of language – manage to get their meanings across from one situation to the next. Thus it promises a welcome alternative to what many thinkers of a 'post-analytic' persuasion (Schiffer and Davidson among them) now regard as the manifest shortcomings of a formalized or truth-theoretic approach. But there would seem little point in holding such a theory – even a minimalist theory-to-end-all-theories – if it can claim only the negative merit of placing no extra (self-induced) burdens on philosophy of language that might be avoided by simply backing down on all the main points at issue. For in so far as that project has any purpose at all it is to offer an adequate *theoretical* account of what speakers and interpreters normally do without wishing or needing to theorize.

This was exactly Davidson's argument in his writings of the early- to mid-1970s, whatever the ambivalence that can now be discerned – with benefit of hindsight – in his use of certain terms like 'truth', 'meaning' and 'language'. It also corresponds to the position that Schiffer adopted in his previous book *Meaning* (1972), a position laid out for clinical inspection in the litany of wanhope – the catalogue of now abandoned ideas – rehearsed in the Preface to *Remnants of Meaning*. But in both cases what we are here invited to contemplate is not just the negative outcome of a specialized intra-philosophical dispute concerning the scope and limits of analysis. Rather it is the much more decisive failure to provide any adequate account of what goes on in even our most basic – let alone our more complex, creative or challenging – acts of linguistic communication. And if it is asked by what standards the project is deemed to have failed then, again, the answer is not (or not only) by the standards of that earlier formalized or truth-theoretical approach which of course neither Schiffer nor Davidson would now accept as providing all the relevant criteria. For there is a further, more damaging loss to be reckoned with in this current retreat from the philosophic high ground, or this scramble to vacate what is now perceived – not without reason – as an exposed position too far above the level of everyday practical-communicative grasp. It is the danger of swinging right across in the opposite direction, that is, toward a minimalist semantics (or, in Schiffer's version, a 'no-theory theory of meaning') which would simply have nothing relevant to say about language, meaning, and interpretation.

The criterion of relevance or adequacy here is one that derives not so much from the formal-semanticist programme as from the tacit knowledge, among speakers and interpreters, that there is more to the process of linguistic 'uptake' than can possibly be accounted for on any such minimalist terms. In short, it is a standard which necessarily involves some appeal to language-user's intuition as the measure of theoretical success. But this is not to argue – with the later Davidson – that 'wit, luck, and wisdom' are all we have to go on, whether as everyday speakers/interpreters of language or as theorists who can likewise avoid a lot of trouble by adopting the minimalist line, shedding all that surplus conceptual baggage, and simply taking each utterance as it comes. For it is not just the theorists and linguistic philosophers who are apt to emerge from this process with a sharply diminished sense of intellectual or professional self-esteem. More than that, it entails a drastic under-rating of the kinds and degrees of complexity – the tacit dimensions of grammatical and logico-semantic grasp – that are merely hinted at in the straightforward appeal to language-user's intuition.

Schiffer registers a certain unease on this account when he re-marks toward the end of his book that he would not wish to think of his career as devoted solely to the therapeutic task of 'show[ing] the fly the way out of the fly-bottle' (*Remnants*, p. 271). To take such a Wittgensteinian view would amount (one gathers) to an outright counsel of despair rather than a fairly honourable defeat on terms of his own choosing. After all, Schiffer's project – early *and* late – belongs very much to that analytic strain within philosophy of language which has held out against the standard Wittgensteinian appeal to 'language-games', or cultural 'forms of life' as the final source of wisdom in all such matters. If this were indeed the end of all his travails then Schiffer would have done much better to fall silent than write yet another book on the subject, much of it pitched (as before) at a highly technical level. But by this stage he has travelled so far along the road to disillusion that one cannot im-agine how his project might issue in anything more hopeful or constructive. So it is that Schiffer's story ends up with the 'no-the-ory theory of meaning', an outcome which lacks even Davidson's sanguine (if scarcely convincing) assurance that speakers and inter-preters can cope just as well in the absence of any such resources.

It is worth recalling just how many items of belief – common-sense-intuitive as well more specialized linguistic-philosophical items – have had to be jettisoned in the course of arriving at this defeatist outcome. They include: (1) the theory of *propositional con-tent* which relates meaning to truth-values through an analysis of logico-semantic structures of representation; (2) the idea of beliefs, attitudes, assenting dispositions, and other such intentional (speaker-related) mind-states as bearing upon – and explicable in terms of – those same first-order propositional contents; and (3) the assumption that there must exist entities (objects, realia, natural kinds, classes of object, etc.) that correspond to those various prop-ositions or beliefs. Items (1) and (2) are Schiffer's main targets in the closing chapter of his book. Thus: 'the no-theory theory of meaning has two components: one, the no-theory theory of linguistic repre-sentation, that pertains to language and to meaning in a strict sense;

and another, the no-theory theory of mental representation, that pertains to the intentionality, or content, of propositional attitudes' (*Remnants*, p. 265). In the case of item (3) Schiffer's argument entails the extreme nominalist position that in fact there are no such objects, entities, kinds, or classes except in so far as they are projected onto the world by this or that (language-relative) classifying scheme. Any stronger form of ontological commitment is in his view merely the result of our supposing that propositions and beliefs can have no determinate content – no meaning or truth-value – unless they correspond to the way things are in some real-world ontological domain. But if one denies (like Schiffer) 'that there are any such *things* as meanings', then of course there is no longer this need to populate the world with actually existing objects, entities, kinds and so forth to which those meanings (along with their associated attitudes and beliefs) must somehow be thought to correspond.

It might seem odd to suggest, as I did above, that these technical-sounding arguments go as much against our 'commonsense-intuitive' ideas about linguistic understanding as against the more specialized sorts of preconception that have typified recent linguistic philosophy. After all, it could be said that Davidson and Schiffer have at last come around – after many distractions along the way – to a no-nonsense viewpoint which allows speakers and interpreters to do what they normally (and for the most part successfully) do without introducing unnecessary problems about 'meaning', 'belief, 'propositional content', and the rest. However, this ignores the far more serious difficulties that arise when Davidson and Schiffer push right through with the minimalist-semantic approach or the 'no-theory theory of meaning'. For it involves rejecting a whole raft of ideas, beliefs, and assumptions which are everywhere bound up with our everyday (commonsense-intuitive) modes of speech and behaviour. Thus in Fodor's aptly chosen words:

If you ask the man on the Clapham omnibus what precisely he is doing there, he will tell you a story along the following lines: 'I wanted to get home (to work, to Auntie's) and I have reason to believe that there – or somewhere near it – is where this omnibus is going'. It is, in short, untendentious that people regularly account for their voluntary behaviour by citing beliefs and desires that they entertain; and that, if their behaviour is challenged, they regularly defend it by maintaining the rationality of the beliefs

('Because it *says* it's going to Clapham') and the probity of the desires ('Because it's *nice* visiting Auntie').[21]

This connects with the argument – going back to Aristotle – that any reasoning on issues theoretical or practical must involve some appeal to propositional content, that is to say, some grasp of what follows from what as a matter of valid inference. Thus in the case of certain 'practical syllogisms' – that is, those pertaining to choices of conduct – the most appropriate outcome may take the form of an *act or decision* that follows from the premises, rather than a statement (or a further proposition) consistent with them. But this argument also works in reverse. For our capacity to draw such inferences (and act upon them) must presuppose *first* that we have correctly understood their propositional content – their meaning, structure, and order of logical entailment-relations – and *second*, that we are able to arrive at beliefs (whether for ourselves or through the process of assigning them to others) in terms of assenting or dissenting attitudes to that same propositional content.

It is just this twofold supposition that Schiffer sets out to undermine by arguing that 'meanings are not things', that they possess no semantically or truth-theoretically determinable content, and therefore that they cannot be objects of belief as required by the relational theory of propositional attitudes. Indeed, when described like this, the theory does sound both forbiddingly technical and disablingly remote from our everyday practices of meaning- and belief-attribution. But such a response would miss the point in at least two respects. One is the requirement that philosophy should provide a more perspicuous (logically accountable) analysis of meanings, motives, beliefs, reasons, and grounds for action than we might very often have need of (or time for) in the course of our everyday lives. Schiffer's analytical training shows through in his continuing to honour this requirement, even when he sets out to dismantle most of the aims and ambitions to which it has given rise over the past five decades and more. But there is also the point – as I have argued above – that in retreating from one, highly specialized mode of complex formal analysis (that is, the Tarskian truth-theoretic approach) philosophers may find themselves driven to adopt a minimalist semantics, or a 'no-theory theory of meaning', which leads them to ignore or summarily reject a whole range of alternative resources. Hence the various counter-movements that have sprung up very largely – it is tempting to suggest

– by way of a defensive over-reaction to the perceived failures (or excessive ambitions) of philosophy in the analytic mode. Chief among them is the recourse to 'ordinary language', to 'language-games' or cultural 'forms of life' as an escape-route from problems which are held to arise only on account of our seeking explanations or conceptual clarity where in truth there is nothing to be explained or clarified – nothing, that is, save our own needless 'bewitchment by language' in having raised such questions to begin with.[22]

Nor is this reactive movement by any means confined to philosophy of language after the manner of late Wittgenstein or Austin. It is also clearly visible in ethics, political theory and the social sciences, most often where these have taken the turn toward a communitarian way of thinking which likewise rejects the delusory appeal to anything – whether 'reasons', 'grounds', or 'principles' – beyond what is sanctioned by tradition or by customary usage in various contexts of enquiry.[23] For the effect, once again, is to flag certain areas strictly off-bounds not only for philosophers and theorists but also – it is strongly implied – for 'ordinary' language-users or moral-political agents whose words, thoughts and actions can only make sense against this presumed background of tacitly binding social-communicative norms. In other words, it tends toward a low estimate of the scope for self-knowledge and rational understanding, as well as for the critical assessment and analysis of other people's actions and beliefs.

CHARITY, TRUTH, AND INTERPRETATION

We can get some useful bearings here by contrasting Davidson's generalized 'principle of charity' with David Papineau's more discriminating 'principle of humanity' as rival accounts of what transpires in the process of construing meanings and intentions.[24] For there is a marked kinship between the Wittgensteinian doctrine of meaning-as-use and the upshot of Davidson's approach. This is manifest in the way that 'charity' is transformed – by the time of 'A Nice Derangement' – into something that functions less like a principle for optimizing true beliefs and more like a shorthand label of convenience for minimalist semantics or the no-theory theory of meaning. Of course it may be said that Wittgenstein stresses the communal (socially-sanctioned) aspects of language-use where Davidson appears to go in quite the opposite direction. That is to say, he espouses a radically occasionalist account that would take each utterance as it comes and cut out the appeal to any background context of shared linguistic knowledge or interpretative grasp. However, what unites them – together with Schiffer – is the belief that *theories* can play no useful or constructive role in this process, whether formalized (higher-level) theories of the kind that linguists and philosophers typically come up with, or theories imputed to natural-language speakers and interpreters in order to explain how they do things with words.

For Papineau, conversely, there is no good reason – in theory or in practice – to place such limits on the human capacity for figuring out meanings and intentions through a complex process of belief-attribution. Thus he accepts the Davidsonian Principle of Charity to this extent at least: that we cannot make a start in interpreting other people's utterances unless we are prepared to count them 'right in most matters', or possessed (like us) of the requisite procedures – the powers of rational inference – to arrive at true beliefs and assign correct interpretations. However it is always possible that they (or we) may turn out to have arrived at some false belief, or some mistaken belief-attribution, which gets in the way of this otherwise perfect, unimpeded communion of minds. Indeed – and this is Papineau's stronger claim – we should have no grounds for making judgements of truth (or ascriptions of justified belief) were it not for the standing possibility of error at some stage in the communicative process. Thus the speaker may perhaps be mistaken in believing this or that to be the case, or again, may produce some inappropriate (semantically or grammatically deviant) form of words that fails to make sense in context. The interpreter will then be faced with a choice between the different kinds of optimizing strategy on offer. These include (1) counting the speaker linguistically competent but wrong as regards some particular item of belief; (2) reinterpreting her or his words so as to make the belief come out true with allowance for *just the requisite degree* of assumed linguistic deviation; and (3) applying the Principle of Charity in its wholesale form, that is, as a maxim which holds: 'always interpret on the generous assumption that if people seem to utter false beliefs (or talk nonsense) then they must have something quite different in mind'. From which it follows that *anything they say* is sure to be true (or at any rate meaningful in context) since we can always appeal direct to utterer's intention and simply disregard any obstacles along the way.

However, if we operate on (3) then we can have no means of distinguishing between instances of (1) and (2), that is to say, between cases of erroneous belief and cases where the better – more charitable – option is to put the problem down to some localized failure of linguistic-communicative grasp. Worse still: we may conclude (like Davidson) that such problems are by no means exceptional but in fact provide a fair sample of what goes on in the everyday business of interpreting speech-acts from one situation to the next.[25] In which case, we might as well abandon the idea that correct interpretation has anything whatsoever to do with understanding a speaker's meanings, attitudes, intentions or beliefs through the best (most rational and at the same time most generous) construal of their words in context. For we shall then be reduced to the desperate expedient – as with Sheridan's Mrs Malaprop – of just ignoring the sense of what they say and imputing a wholly conjectural 'utterer's meaning' that bears no relation to linguistic meaning nor indeed to any process of belief-attribution that would seek some reason – some linguistic, motivational, or causal-explanatory account – for their tendency to say such things. In effect this leaves both parties (speaker and interpreter) possessed of nothing more than 'wit, luck, and wisdom' plus a shared sense of how little they share in the way of linguistic and interpretative resources.

It is in order to avoid this unfortunate upshot that Papineau proposes the switch from 'charity' to 'humanity', that is, from an empty (because all-purpose) principle of optimizing true beliefs to a principle which holds that speakers will *most likely* have something truthful, intelligible, or meaningful to say *unless* we have good reason to suspect otherwise.[26] And in the latter case we shall not feel obliged, like Davidson, to operate the kind of wholesale interpretative discount-theory which extends them the benefit of the doubt on condition that their utterances, meanings and beliefs are subject to radical reconstrual. For on Davidson's account – as in so many treatments of this issue after Quine – there is an odd tendency to regard all contexts of linguistic-communicative behaviour as resembling the predicament of a 'field anthropologist' (or 'radical interpreter') confronted with a 'native informant' whose every word and gesture must be taken as belonging to some wholly unintelligible language-game, cultural 'form of life', or whatever.[27] In 'On the Very Idea of a Conceptual Scheme' Davidson responds to this challenge by deploying his generalized Principle of Charity, that is, his argument that we just *could not possibly* be wrong in the majority of our meaning- and belief-attributions. For if meaning is given by truth-conditions, and if it is simply *unintelligible* (as Davidson contends) that either party should be 'massively in error' with regard to their most basic beliefs, then there is no need to follow Quine all the way to his sceptical-relativist conclusions.[28] However, as we have seen, this argument provides nothing more than a formal or abstract assurance that communication is possible across and despite divergences of 'conceptual scheme'. For it fails to give any adequate account of those logico-semantic operations which explain how it is that speakers and interpreters actually manage such a complex business as figuring out meanings, intentions and beliefs on the basis of such logically 'primitive' notions as truth and falsehood. Hence Davidson's retreat to a minimalist semantics, to a no-theory theory of meaning where the erstwhile 'native informant' has now been replaced by Mrs Malaprop, a speaker whose aphasic ramblings are so far off-beam as to stretch charity to the limit and beyond.

At this point, as Papineau remarks, 'the Davidsonian argument against divergence of belief has collapsed altogether'.[29] What started out as a theory for optimizing the prospects of shared communicative grasp across otherwise sizable differences of language and culture has now ended up with the flat denial that there is any such thing as (knowing or sharing) a language. Thus '[n]ot only are there no *a priori* limits to divergence of opinion, there seem to be no *a priori* limits to divergence of concepts either. . . . After the initial promise of the "transcendental argument to end all transcendental arguments", this might all seem rather disappointing'.[30] It is worth noting that Papineau's book appeared in 1987, before the publication of 'A Nice Derangement of Epitaphs'. (The latter had been circulating in typescript for some years prior to that but there is no indication that Papineau had read it.) His criticisms thus have an added degree of diagnostic or predictive force, based as they are on a shrewd perception of problems with the Principle of Charity that stand revealed more clearly in Davidson's turn toward a full-fledged minimalist semantics. In any case the alternative, as Papineau sees it, is not far to seek. 'The whole point of the switch to humanity is that on occasion we find it *intelligible* that *in such-and-such circumstances* people are likely to have *false* beliefs.'[31] And again: 'once we switch to humanity, and thereby open the way to intrinsic identifications of belief, we end up allowing divergence of concepts as well as divergence of opinions'.[32] In which case, it might be thought, there is not after all so very much to choose between Humanity and Charity, since this allowance for divergent concepts and beliefs is exactly what Papineau finds Davidson forced to concede despite his attempt to prove otherwise. But with Davidson it produces a swing to the opposite extreme, a divergence-thesis that would place such a gulf between speaker and interpreter as to render their occasional *con*vergence on 'passing theories' an event quite beyond reach of linguistic or rational accountability.

It is for much the same reason that Schiffer is hard put – after the failure (as he sees it) of truth-theoretical semantics in whatever form – to explain how his position now differs significantly from the Wittgensteinian doctrine of meaning-as-use. At any rate one can see why his negative verdict in respect to that programme might be thought to leave him with no other option. Thus it is a fallacy (he argues) to suppose that speakers, interpreters, or agents should be thought of as performing some complex set of operations – some procedural equivalent to a formalized intention-based semantics – whenever they engage in such activities. And this is also the line taken (albeit less fretfully) by thinkers of a Wittgensteinian persuasion – philosophers, ethnographers, cultural historians, advocates of the 'strong programme' in sociology of knowledge – who likewise reject any notion of truth beyond the currency of beliefs-held-true within some given historical or cultural context.[33] On this view we shall come no closer to understanding what speakers mean or what believers believe through some attempted construal of this or that utterance in truth-theoretic terms. Rather their words, their beliefs and intentions either do or do not make sense according to the various implicit criteria that decide what shall count as a meaningful (or veridical) utterance by their own interpretative lights. And since there are as many 'language- games' as 'forms of life' – or as many relevant contexts of belief as beliefs that make sense in those same contexts – we shall surely be mistaken, so the argument runs, if we think to distinguish between them in respect of validity or truth.

This doctrine has lately come in for much criticism and I shall not rehearse the main arguments in any detail.[34] Sufficient to say that it makes no allowance for the complexities of motive, meaning and intent that characterize many human predicaments both in the ethical and the linguistic-interpretative sphere. Conflict-resolution becomes simply a matter of invoking the relevant communal norms, rather than a process of thinking through – and hence coming to appreciate more fully – the sorts of complication involved. At best this produces an attitude of liberal tolerance for the variety of human cultures, values and beliefs, though an attitude that also lies open to the charge of evading significant moral and interpretative issues by adopting a relativist perspective. At worst – as some of its critics have argued, Onora O'Neill among them – it tends to promote a complacent or a narrowly parochial outlook, a belief that such issues can only make sense by criteria 'internal' to the language or culture concerned.[35] This has the twofold disabling consequence: (1) that we are effectively debarred from criticizing others, from holding their beliefs mistaken or their ethical values unjustified (since we thereby demonstrate our own lack of inwardness with the beliefs and values concerned), and (2) that we shall never meet any challenge to our own acculturated habits of thought since again such a challenge would be simply *unintelligible* to us as denizens of our particular linguistic, cultural, or socio-historical lifeworld. For in order to criticize a language-game or a cultural 'form of life' one must take up a certain distance from it, at least to the extent of not being bound to judge and interpret always in accordance with its own evaluative criteria or communal norms. From which it follows, on this internalist account, that one will then not so much be *interpreting* the language and beliefs in question but applying a different (external) set of criteria. What is more, any statement, truth-claim or evaluative judgement within that language must always be construed holistically, that is, as bound up with the entire existing range of social and communicative practices which constitute the cultural life-form concerned. Thus to criticize any one or more of its component beliefs is *de facto* to cut oneself off from the kind of inward (sympathetic) understanding which alone makes it possible to properly interpret those beliefs.

Peter Winch's work is the best-known example of this Wittgensteinian approach to issues in sociology and the human sciences at large.[36] But the extent of its influence can perhaps best be gauged by noting how it has now become a central motif in the writings of Alasdair MacIntyre, a social critic and ethical philosopher whose early book *Against the Self-Images of the Age* contained some of the most cogent criticism of this widespread Wittgensteinian drift in philosophy and the social sciences.[37] MacIntyre would no doubt reject such a claim, since he now regards the whole debate about cultural relativism as one more melancholy symptom of decline in a story whose episodes include every stage of Western (post-Hellenic) thought and culture.[38] On his account it shows that we have entered a late – perhaps terminal – phase in the loss of those shared and substantive ethical beliefs that characterized the conduct of life in the ancient Greek *polis,* and which Aristotle expressed most fully in his doctrine of the virtues. However this commits MacIntyre to the view that such virtues must be relative to some given place and time, some existing set of practices, values and beliefs that provide the necessary background conditions within which they are able to develop and flourish. His own preference is for talk of 'narratives' – stories that people and polities can live by – rather than the standard Wittgensteinian appeal to 'language-games' or cultural lifeworlds. What he most admires about Aristotle's doctrine is its capacity to weave together the various aspects of a virtuous life – individual, domestic, philosophical, ethical, civic, socio-political, and so forth – in the form of a model narrative to which all citizens can aspire. Such a conception would involve none of those vexing dualisms (as between 'private' and 'public' duty or pleasure and moral obligation) which have been the bane of ethical philosophy at least since Kant.

All the same it is hard to square MacIntyre's praise for these distinctively ancient-Greek virtues – or for the long-lost 'organic' ethical community to which they bore witness – with his hope (albeit a forlorn hope) that we might yet regain something like that sense of individual and communal purpose. And there are other, more conspicuous problems with his argument, among them the fact that Aristotelian ethics (and contemporary ideas of what counted as social justice) applied only to free-born Greek male citizens and licensed all manner of prejudice, oppression or brutality in the case of women and slaves. I shall not belabour the point here except to remark that it offers a particularly vivid instance of the kinds of difficulty that moral theorists get into when they seek – like MacIntyre – to bypass the problems of complex ethical and interpretative judgement by invoking some version of the forms-of-life doctrine, or the appeal to 'criteria' presumed to exist within this or that cultural community. For there is no good reason (ingrained scepticism aside) to suppose that moral agents, speakers and interpreters can only make sense of themselves – or others – on the terms and membership-conditions laid down by this highly restrictive ordinance.

ALTERNATIVE RESOURCES: A REALIST APPROACH

I hope it will be evident by now why I have brought together such a range of issues from recent philosophical debate. For they all have to do with the central question as to just how far analysis can take us beyond that point where, in Wittgenstein's view, 'explanations' and 'reasons' at last have an end and interpretation can only proceed on a basis of shared (communally sanctioned) meanings, values, and beliefs. In ethics and philosophy of action this issue works out as a conflict of views between those, like Winch and MacIntyre, who adopt some version of the communitarian standpoint and those – like O'Neill – who argue against it on the grounds that it treats moral agents as creatures of acculturated habit and custom, unable to confront any serious challenge to the values that prevail in their own society. On this account, as O'Neill puts it, 'moral reasoning presupposes shared moral traditions and practices', so that 'only within such a context can moral discourse about examples take place, and questioning of the shared framework of moral practices is not possible'.[39] Moreover, 'even those Wittgensteinian writers who reject relativist readings of Wittgenstein do not offer an account of moral practice and decision that goes beyond the practice-based conception of ethical decision offered by relativist writers'.[40] That MacIntyre's work should fall under this general description is all the more ironic given the title of that earlier book (*Against the Self-Images of the Age*) in which he offered a cogent critique of Winch for much the same reasons as O'Neill. One might remark also that his reading of Aristotle tends to play down precisely those aspects of Aristotelian ethics – like the relation of moral reasoning, *via* the practical syllogism, to other modes of rational inference – where the emphasis falls (as I noted above) on the capacity of human agents to think and judge for themselves. For this capacity is allowed no scope – or a drastically diminished scope – by any argument that would treat moral issues, choices and decisions as intelligible only by the shared criteria pertaining to some given communal life-form.

The same applies to philosophy of language and interpretation-theory as construed on the minimalist-semantic account proposed by (the later) Schiffer and by Davidson in 'A Nice Derangement'. For here likewise there is a full-scale retreat from the notion that speakers and interpreters might share not only 'a language' but also, more specifically, the means to produce and understand novel sentences of a language through the kind of logico-semantic grasp which goes well beyond any straightforward appeal to accultur-ated habits of usage. In Davidson's case this retreat takes the form of first proposing, then distinctly playing down, a formalized (Tar-skian or truth-theoretical) semantics which turns out not to be of much use in the natural-language or everyday communicative con-text. With Schiffer it involves a painful acknowledgement that nothing remains of his original programme, of 'propositional con-tent', determinate beliefs, intention-based semantics, or other such 'remnants of meaning' analytically construed. And yet, and yet . . . Schiffer cannot quite bring himself to take the ready Wittgenstein-ian route out of all these difficulties. Thus he does raise the question 'whether there is any good reason for or against supposing natural languages to have compositional meaning theories that are not compositional truth-theoretic semantics', or again, 'any positive reason, as there is against compositional truth-theoretic semantics, for supposing that there cannot be correct meaning theories' (*Rem-nants*, p. 233). If there were such a theory – he is ready to concede – then it might hold out a promising alternative to Schiffer's hard-put agnostic conclusion. However, having ventured the hypothesis and viewed it from various angles, he decides that this is not after all a genuine option.

For if knowledge of a theory really sufficed to enable one to understand utterances in the language that the theory was about, then knowledge of the theory would enable one to know the saying potential of the sentences of the language. But knowing this, for declarative sentences, would be tantamount to knowing the truth conditions of utterances that could have them; and how could such a theory fail to be, in the sense stipulated, a composi-tional semantics? (Ibid.)

In Schiffer's view this would make it necessarily a non-starter since no such formalized (truth-theoretic) account can possibly come up with the desired correlation between 'propositional con-tent' on the one hand and beliefs, intentions, or utterer's meaning on the other.

Still one may doubt whether Schiffer is justified in this claim to have conducted a critical review of *all possible* alternative truth-based semantic theories and moreover to have shown that they each fall prey to the same (presumptively fatal) range of objections and counter-arguments. For he passes too quickly over several such candidate theories that might offer a cure for his far-gone sceptical malaise. One is the causal account of naming, necessity and natural kinds which receives its most elaborate – and convincing – treatment in the work of Saul Kripke and Hilary Putnam.[41] According to this account reference is fixed through an initial act of designation which picks out an object (prototypically a natural kind such as *gold* or *water*) and is thenceforth applied to any sample instance that meets the relevant identification criteria. These in turn may be refined or specified with greater accuracy through some advance in scientific knowledge, as for instance when 'gold' is assigned the atomic number 79 or when 'water' is discovered to have the molecular constitution H_2O.

One advantage of the causal theory is that it avoids some of the counter-intuitive consequences that result from a purely descriptivist account. On this latter view – the Fregean thesis that 'sense determines reference' – what enables us to pick out specimens of 'gold' or 'water' is our knowledge of the relevant identification-criteria as given by the meaning of those terms.[42] Thus *gold* is whatever answers the description 'yellow', 'ductile', 'dissolvable in *aqua regia*', or 'having the atomic number 79', and *water* the substance likewise identified as 'liquid', 'colourless', 'odourless', or by its freezing-point, boiling-point, molecular constitution, and other such descriptive attributes. The problem with this theory, as Kripke points out, is that any new discovery concerning such substances – or revision to the range of accepted criteria – will entail a corresponding shift of attributes, such that *water* is no longer 'water' and *gold* no longer 'gold'. (In the case of proper names the argument would run: if we discover Aristotle not to have been the pupil of Plato, tutor of Alexander, author of the *Posterior Analytics*, etc., then it follows of necessity that *Aristotle* was not in fact 'Aristotle'.) Since these conclusions are counter-intuitive to the point of manifest absurdity they must be based on a false understanding of the relation between sense and reference. Hence the alternative solution proposed by Kripke and Putnam: that reference is fixed once and for all by an original act of designative naming. Then one can account for theory-change without being driven to Quinean conclusions about wholesale 'ontological relativity' or Kuhnian ideas about radical meaning-variance across and between paradigms. Any shift in the meaning of terms – or in the range of criteria for their valid application – must still be taken as referring to *just that*

entity, rather than some wholly other notional entity whose existence can only be a product of semantic definition.

This is one of the alternatives that Schiffer rejects, not so much out
of hand (since he spends several pages discussing Kripke's argument) as through a strong preconception that it *must* fall prey to
some one or more of the sceptical arguments against any form of
truth-theoretic or intention-based semantics. For clearly if those
proposals held up then there could be no reason – at any rate no
philosophically compelling reason – to follow Schiffer all the way
to his defeatist conclusion. Thus it might more plausibly be argued:
(1) that propositions have meaning (or a determinate truth-value)
in so far as they assign valid predicates to referring expressions
whose reference is fixed in the manner that Kripke describes; (2)
that beliefs have content in so far as they involve some specifiable
attitude (e.g., that of holding-true) toward such propositions; and
(3) that one can thereby assess those beliefs not only with respect to
their truth or falsehood but also as concerns their intelligibility as
the kinds of attitude that speakers might be expected to hold given
what we know of their cultural background, information sources,
cognitive priorities, and so on. If this all sounds very much like
Papineau's 'Principle of Humanity', as summarized above, then the
reason is not far to seek. For in both cases the argument is that we
should make sense of utterances and beliefs through an apt combination of (qualified) charity, acquired linguistic knowledge, and –
where needed – inference to the best (most adequate or convincing)
causal explanation. Only thus can we reconcile the two main demands upon any such theory: that it optimize the content of utterer's beliefs while not (as with Davidson) extending charity to the
point where it gets altogether out of touch with criteria of truth and
falsehood or warranted assertability.

Of course these are not the kinds of issue directly addressed by
Kripke or by Putnam in his work on the causal theory of reference.
Their aim is to demonstrate the order of *metaphysical* necessity that
determines the relation between proper names (or natural-kind
terms) and the objects to which those names refer. In this way they
seek to avoid all the problems thrown up by descriptivist theories
and also by the whole tradition of epistemological enquiry going
back to Descartes, Hume, and Kant. Hence Kripke's most striking
and widely debated claim: that such necessity is *a posteriori* in
character, analogous to the kinds of empirical discovery-procedure
exemplified by the natural sciences, rather than resting on a

dubious appeal to *a priori* (epistemological) grounds of knowledge.[43] In which case it might be thought that Kripkean semantics is restricted by design to just those aspects of language that fall within the remit of a purely extensional analysis, as opposed to those other (intensional) aspects – of belief, holding-true, holding-false, qualified acceptance, and so forth – which remain in some degree opaque to any such analysis. However there are also instances – discussed by various commentators, Gareth Evans among them – that clearly call for a joint application of the causal and descriptivist theories. For as Evans points out there may often arise situations where the causal theory (that reference is 'fixed' by some original act of naming) cannot go far toward explaining how it is that subsequent speakers may apply the same term in a perfectly appropriate manner but without any possible – direct or circuitous – linking-back to that inaugural act through the causal chain of transmission.[44]

In some cases, moreover, it may even be a *necessary condition* for correct reference (or true belief) that the chain should at some point be broken and the speaker apply criteria at variance with those of his or her information-source. Thus: 'suppose on a T.V. quiz programme I am asked to name a capital city and I say "Kingston is the capital of Jamaica", I should want to say that I had said something strictly and literally true even though it turns out that the man from whom I had picked up this scrap of information was actually referring to Kingston-upon-Thames and making a racist observation'.[45] And again:

> Amongst the Wagera Indians, 'newly born children receive the names of deceased members of their family according to fixed rules . . . the first born takes on the name of the paternal grandfather, the second that of the father's eldest brother, the third that of the maternal grandfather.' In these and other situations (names for streets in U.S. cities, etc.), a knowledgeable speaker may excogitate a name and use it to denote some item which bears it without any causal connection whatever with the use by others of that name.[46]

This is not to deny that the causal theory of reference helps to make sense of a good many other cases where the descriptivist theory turns out to have counter-intuitive or absurd consequences. Furthermore, as Evans remarks, it establishes a link between these issues in philosophical semantics and other areas – epistemology, cognitive psychology, and philosophy of mind – whose 'major concepts . . . have causality embedded within them'.[47] Thus the Kripkean theory goes at least some way toward explaining how reference might get fixed – and meanings or beliefs have veridical content – by the same sort of process that typically occurs in cases of direct (undistorted) perceptual acquaintance. 'Seeing and know-ing are both good examples', Evans writes, since they can both be construed – like referring – as achievement-terms whose meaning *necessarily* entails some justifying causal component. After all,

> [t]he absurdity in supposing that the denotation of our contem-porary use of the name 'Aristotle' could be some unknown (n.b.) item whose doings are causally isolated from our body of infor-mation is strictly parallel to the absurdity of supposing that one might be seeing something one has no causal contact with solely upon the ground that there is a splendid match between object and visual impression.[48]

Still there are cases that cannot be accounted for by any purely causal theory, that is, any theory which finds no room for descrip-tivist criteria, intensional meanings, belief-states, or epistemologi-cal (as distinct from causal-epistemic) grounds. These include not only instances of *correct but misattributed* belief, such as the two cited above, but also instances of *justified true belief* where the causal component is insufficient to provide all the necessary background knowledge.[49] For in many such cases it will be more plausible to suppose that speakers can have used the term rightly – in accordance with its proper meaning, reference, and truth-conditions – only in so far as they (or the language-community to which they belong) have adequate criteria whereby to judge of its correct application.

Thus, as Evans ecumenically sees it, there is comfort to be had for both rival camps in the current, sharply polarized dispute between descriptivist and causal theories of reference. The descriptivist can still argue that 'for the most fundamental case of the use of names . . . the idea that their denotation is fixed in a more or less compli-cated way by the associated bodies of information that one could cull from the users of the name turns out not to be so wide of the mark'.[50] That is, there must always be room for criteria of correct (or incorrect) usage and of true (or false) belief that involve on the one hand the shared possession of a language, along with its resources for meaning- and belief-attribution, and on the other hand a grasp of the relevant conditions for assigning truth-values to those meanings and beliefs in accordance with the current best state of knowledge as concerns such candidate items. For the descriptivist is right to this extent at least: that we could be in no position to pick out objects – to fix their reference the first time around or refer to them again in subsequent contexts – if they did not fall under some description (some set of identifying criteria) which served to individuate both the objects in question and any valid belief concerning them. But the causal theorist can also take heart from this account, Evans thinks, 'incorporating as it does his insight about the importance of causality into a central position'.[51] Thus it remains crucial that our paradigm cases of true belief – and of the justified inferences to speaker's meaning based upon them – should take due stock of such causal factors. This we can do, moreover, while acknowledging the role of descriptive criteria as an additional (and sometimes a needed alternative) resource for picking out the referent in question. In fact the only theorists who can derive no comfort from Evans's various examples are 'those who, ignoring Kripke's explicit remarks to the contrary, suppose that the Causal Theory could provide them with a totally *non-intensional* answer to the problems posed by names'.[52] And in their case, he concludes, the solution is simply to adopt a more flexible or less one-sided approach.

My point in all this is that one can – *contra* Schiffer – combine a sturdily realist account of referring expressions with an intensional (truth-theoretic or proposition-based) mode of semantic analysis. If Schiffer is driven to deny both claims it is not so much by reason of their inherent implausibility as because he construes them – like Evans's comfortless third party – in misleadingly narrow and exclusive terms. One can see how this sceptical strategy operates in Schiffer's treatment of the sentence 'Tanya believes that Gustav is a dog', a sentence which on his account contains no determinate proposition that could ever be singled out as the object of Tanya's belief. It could have such a content, he argues, 'only if there really exists a non-pleonastic, genuinely language-independent property of being a dog' (*Remnants*, p. 71). However that property cannot exist since it would surely be absurd to suppose 'that Gustav, in addition to having a certain appearance, demeanour, morphology, ancestry, genetic structure, and so on, also has, over and above all these properties, the quite distinct, primitive, and irreducible property of being a dog' (ibid.).

Still one might ask what more could possibly be needed in the way of identifying criteria. For the properties as listed can surely suffice to specify the meaning and truth-conditions for the sentence 'Gustav is a dog', and also to provide propositional content for Tanya's belief. According to Schiffer these requirements could never be satisfied by this or *any possible* list of attributes, no matter how long or elaborate. That is, there would always be lacking some essence of (or deep further fact about) doghood whose presence was requisite in order for Tanya's belief to have a well-defined content, but which no amount of detailed specification in the above manner could serve to fix or individuate. For 'none of Gustav's properties, however complex, is such that, necessarily, a thing is a dog if and only if it has that property' (p. 71). This follows from Schiffer's attitude of across-the-board nominalism, his refusal to concede *either* any version of the natural-kinds argument (that objects can be picked out in virtue of their possessing certain physically salient properties), *or* any version of the descriptivist case (that reference can be secured through an adequate grasp of the relevant identifying terms and criteria). All the same '[t]he existence of doghood is, for me, of little moment; the issue is raised now for its foreshadowing effect with respect to the issue, later to be pressed, of the existence of objective, language-independent belief-properties, properties such as being a belief that Gustav is a dog' (ibid.). But if the argument fails at this point – if there *are* indeed adequate grounds for Tanya's belief which can be specified in referential, truth-theoretic, propositional, and intention-based terms – then there would seem no reason (prejudice aside) for Schiffer to adopt such a sceptical view.

That he does so none the less – to the point of destroying all his previously held convictions, along with just about every tenet of recent philosophical semantics and, not least, the 'folk-psychological' (commonsense-intuitive) attitude on matters of meaning, truth and belief – is enough to indicate that something has gone seriously awry with Schiffer's approach to these issues. For such scepticism can appear in some degree plausible only if one takes it, like Schiffer, that analytic philosophy is now confronted *at every turn* with problems and aporias that permit of no escape save by abandoning the entire project and starting out afresh on a minimalist tack relieved of all that surplus conceptual baggage. This attitude comes out most clearly in a passage where Schiffer attacks the 'impulse toward analysis' as the great besetting vice of present-day philosophy in general, and philosophical semantics in particular. The impulse is a strong one, he acknowledges, 'and probably connected with an impulse toward an atomistic metaphysics, a metaphysics that insists on seeing all facts as either atomic or else built up out of atomic facts in recursively specifiable ways' (*Remnants*, p. 263). All the same it is an impulse that can and should be resisted. Thus the failure of intention-based semantics is to this extent a salutary failure, 'just one more example in the unpromising history of philosophical analysis, the search for reductions in the form of interesting necessary and sufficient conditions' (ibid.). For in Schiffer's view it is the two main items of that programme – the truth-theoretic account of propositional meaning and the idea of propositions as objects or contents of belief – which have now collapsed and taken the rest of it along with them. 'It is fortunate', he concludes, 'that scepticism about such undertakings is now the prevailing conviction among philosophers' (ibid.).

However, one may doubt that this conviction is either so widespread or – more importantly – the upshot of any such exhaustive, rigorous and critical review of the field. In philosophical semantics (Kripke, early Putnam, Evans, Donnellan) and in epistemology and philosophy of science (early Putnam again, along with Harré, Bhaskar, Salmon, and others) there is a strong counter-movement toward causal realism which would offer just the kind of alternative that Schiffer so doggedly and self- defeatingly rules out.[53] What this also makes clear is the way that Schiffer, like Davidson, has pinned his faith to one erstwhile candidate-theory – a truth-based semantics in the formalized Tarskian mode – whose failure to come up with the promised goods then precipitates either (in Schiffer's case) a full-scale crisis of philosophic nerve or (with Davidson) the sanguine but just as problematical retreat to semantic minimalism and the no-theory theory of meaning. For this can now be seen as a blind alley that was sure to produce such sceptical conclusions in those who – again like Schiffer and Davidson – can see no use for any formalized theory unless it can be shown to work for natural languages.

The alternative works out, I have argued, as a combination of causal realism with respect to objects, referents, and their properties and an intentionalist (belief-based) propositional semantics with respect to 'opaque' (speaker-relative) contexts of utterance. This latter would be one that succeeded in avoiding the high Tarskian road which leads, *via* a formal (meta-linguistic) account of meaning and truth, to its end-point in Schiffer's and other such variants of the 'post-analytical' trend. That is to say, it would locate the truth-conditions for valid and meaningful utterance in logico-semantic structures internal to natural language and not – as with the Tarskian approach – applied (so to speak) from outside and above. For it is just this presupposed distinction of levels between object-language and metalanguage that leads Schiffer to his defeat-ist verdict: namely, that *no* kind of truth-theoretic approach can overcome the problems which *inevitably* arise in attempting to as-sign determinate content to first-order (natural-language) meanings, intentions, and beliefs. What is needed, in short, is a semantic theory that satisfies the following desiderata. (1) It should explain – or at least make adequate allowance for – the role of various causal factors in fixing reference and establishing the range of humanly intelligible belief- attributions. (2) It should provide some adequate account of how speakers and interpreters (not to mention theorists) can do what Schiffer thinks utterly impossible, that is, come up with a 'correct meaning theory' from one natural-language context to the next. This in turn requires (3) that the sought-for account will relate semantic content to belief-content, or meaning as given by analysis of the relevant truth-conditions to meaning as construed with reference to a speaker's intentions, propositional attitudes, assenting or dissenting dispositions, and so forth. (By the by: lest it be said that this involves a confusion between intensional and intentional modes of analysis I should perhaps add that most efforts to make that distinction stick have ended up in conspicuous failure.)[54] Moreover (4), the theory will need to explain – and more convincingly than Davidson – how language-users manage to converge on a shared interpretation that implicitly involves all the above components of an adequate com-municative grasp. For if Schiffer is right about one thing at least it is his fretful sense that the sceptical 'solution' cannot come close to describing or explaining what makes such convergence possible.

4

Complex Words, Natural Kinds, and the Justification of Belief

ANALYSIS AND ITS DISCONTENTS

In this chapter I shall resume my case that philosophers – more specifically, workers in the field of philosophical semantics – have a great deal to learn from William Empson's books *Seven Types of Ambiguity* (1930) and *The Structure of Complex Words* (1951). I shall also put the case (with an eye to certain recent 'post-analytic' developments) that Empson provides a much-needed alternative approach to issues of meaning, truth and interpretation. In particular he points a way forward from the impasse arrived at by theorists like Donald Davidson, those who espouse a minimalist semantics – or a 'no-theory theory of meaning' – as an escape-route from the project of old-style analytic philosophy.[1] For it is widely held that this project has stalled on the sheer impossibility of producing a truth-based compositional semantics for natural language, that is, some means of individuating beliefs and assigning truth-(or false-hood-) values while respecting the need to interpret speakers on a Davidsonian 'Principle of Charity' that maximises mutual understanding.[2]

Stephen Schiffer, as we have seen, takes the gloomiest view in his book *Fragments of Meaning*.[3] Having started out firmly committed to the programme of Intention-Based Semantics (IBS for short) Schiffer now despairs of carrying that programme through.[4] For there is, he argues, no adequate or non- circular method for determining (1) the propositional content of beliefs; (2) the reference (or extensional scope) of their various constituent terms; (3) their compositional structure expressed in logico-semantic form; and (4) speaker's or utterance-meaning, the latter construed as a relation that obtains between various attitudes (primarily that of holding-true) and belief-content as specified by items (1) to (3) above. To make his point, Schiffer instances the sentence 'Tanya believes that Gustav is a dog' and then goes on, with a certain grim relish, to show what unsuspected difficulties arise in the process of assigning it an adequate interpretation.[5] These include the problems of fixing a referent (species or natural kind) for the word 'dog'; establishing its logico-semantic role as a constituent of the object-sentence ('Gustav is a dog'); and defining the conditions under which Tanya's belief can properly be said to have determinate content and hence a decidable truth-value. Worse still, there is the problem of opaque (intensional) contexts and how to hook them up with an extensionalist account of the satisfaction-conditions for veridical utterance. On each of these counts it is Schiffer's pyrrhic contention that analysis has failed to produce a theory – an intention-based compositional theory – that would come anywhere near to delivering the promised goods.

Schiffer is doubtful, at the end of all this, whether his attitude could fairly be characterized as 'defeatist' or 'despairing'. While rejecting those epithets he is hard put to say why they should not apply. At any rate he lacks Davidson's breezy assurance that 'wit, luck, and wisdom' – plus the occasional 'passing theory' – are all that is required for communicative uptake in the absence of anything that could properly be described as knowing, possessing or sharing a language.[6] Nevertheless there is a sense in which Davidson and Schiffer have travelled the same road of increasing disenchantment with the prospects once offered by a truth-based theory of meaning and interpretation. Thus in Davidson it takes the form of certain well-known throwaway pronouncements ('there is no such thing as a language') coupled with the idea that we can best get along – from one speech-occurrence to the next – on an *ad hoc* mixture of charity, guesswork, and sheer *ad hockery*.

So it is that he can treat Sheridan's Mrs Malaprop as a set-piece example of what goes on when we encounter some novel or hitherto unmet-with item of usage.[7] For, according to Davidson, there is a difference only of degree between the sorts of allowance we learn to make in cases of chronic malapropism and the sorts of localized adjustment that are always required in order to construct a 'passing theory' for this or that particular utterance in context. His point – to recapitulate briefly – is that *every* such utterance will need interpreting with due regard to certain specific (uniquely occurrent) factors such as social setting, idiolectal variation, peculiarities of tone, shared background knowledge, or circumstantial cues and clues. In which case there is no point trying to come up with a generalized account (or 'prior theory') of what it takes for speaker and interpreter to converge on a shared understanding of the utterance in question. For it must always be subject to a law of sharply diminishing returns, that is to say, a situation such that the theory's usefulness for practical purposes varies inversely with the scale or scope of its proposed application. Thus it is in the very nature of a passing theory to work just once – for some particular utterance-in-context – and to lay no claim to an order of generalized (trans-contextual) validity. By the same token, in so far as prior theories have any use at all, they will need to be applied (so to speak) in the default mode or as working hypotheses that may always turn out to offer little or no guidance in matters of communicative uptake. From which it follows – on Davidson's minimalist-semantic account – that prior theories are pretty much redundant and passing theories just a stopgap expedient whose field of application is always, of necessity, 'vanishingly small'.[8]

There could hardly be a greater contrast than that between Empson's prolific insights as a close-reader of texts and the exiguous resources currently on offer from post-analytic philosophers like Schiffer and Davidson. Perhaps it will be said that this comparison is wildly off the point since philosophical semantics and literary criticism are two quite distinct activities with differing standards as to what should count as a valid interpretative insight. However, Empson's *The Structure of Complex Words* is, among other things, a sustained attempt to work out a logico-semantic theory applicable not only to literary texts but to all forms of natural-language utterance.[9] Indeed – as I shall argue – it achieves just what Schiffer despairs of achieving, that is, an intention-based semantics adequate to meet the kinds of objection (on grounds of circularity, redundancy, failure to individuate belief-content, and so forth) that have lately been brought against it. Moreover, his approach helps to indicate just how much is left unaccounted for by Davidson's minimalist-semantic treatment of the prior theory/passing theory distinction. For there is evidence that Davidson had read some parts of the book – in particular its chapter on metaphor – and that his own idea for that distinction came from Empson's broadly parallel treatment of the 'head sense'/'chief sense' pair, the one taken as referring to the dominant (commonplace or standard) sense of a word at any given time, the other to its salient meaning in context as conveyed by various forms of logico-semantic 'equation'.[10] But here, as we shall see, the resemblance comes to an end since Empson exhibits a far greater interest in the way that these senses can always interact so as to generate complex structures of meaning beyond anything dreamt of in Davidson's minimalist theory.

FOR REALISM IN SEMANTICS: EMPSON ON 'DOG'

I shall now return to Empson's *The Structure of Complex Words* since it is here – as I have argued – that these questions receive their most resourceful and promising treatment to date. And since dogs have already come up for discussion it seems only fitting to take his chapter 'The English Dog' as a sample of Empson's approach.[11] Of course there is the likely objection (again) that his interests are so remote from Schiffer's – or from the kind of philosophical semantics that would take for its test-case the protocol sentence 'Tanya believes that Gustav is a dog' – that comparisons are beside the mark. After all, Empson is practising a form of jointly linguistic, cultural, and literary interpretation which does indeed make use of certain logico-semantic formalisms – his so-called 'bits of machinery' – but only in a *post hoc* and piecemeal way, or in order to provide a (somewhat shaky) methodological scaffolding for his otherwise largely intuitive responses. Such was the impression of many early reviewers and also, as far as one can tell, the idea that has stuck among literary critics and those few philosophers (Davidson among them) who have shown any interest in Empson's work. Thus the critics tend to ignore the 'machinery', or consider it a rather tedious distraction from the chapters where it more or less fades into the background and where Empson is free to pursue his bent as an exceptionally gifted close-reader of texts. And in Davidson's case, as we have seen, very little remains of all that complex logico-semantic analysis. What is left at the end of his minimalist road is a tacit appeal to Empson's 'head sense'/'chief sense' distinction – transposed into the idiom of 'prior' *versus* 'passing' theories – and a vague (largely unargued) belief that occasional convergence on passing theories is enough to put speakers and interpreters reliably in touch.

However there is more to *Complex Words* than appears in these skewed or selective readings. If Empson's approach has seemed largely irrelevant to philosophers trained up in the mainstream analytical tradition – or been quarried piecemeal by those, like Davidson, moving in a post-analytic direction – then it might just be that the tradition and its lapsed adherents have missed something of real importance. As I say, the word 'dog' is a useful point of entry for several reasons, not least its cropping up so often in the context of debate about natural kinds, naming, and necessity. Whether dogs can be picked out as objects of belief – or as creatures identified by species-attributes, definite descriptions, individuating properties, and so forth – is not the sort of question that chiefly interests Empson. Still there is nothing self-evidently false or naive (despite Schiffer's hypercultivated qualms) about the common-sense-intuitive assumption that they *can* be thus picked out, along with a good many other such items that figure in our everyday or relatively specialized habits of talk.

The philosophy of biology is one area where this issue – broadly speaking, the quarrel between essentialist and conventionalist taxonomies – has lately provoked a good deal of pertinent debate. In his survey of the field Elliot Sober seeks a middle way between the *extreme realist* position (that there is a 'uniquely correct choice of species-concept' for any given organism) and its *extreme conventionalist* counterpart (that 'every grouping of organisms is just as entitled to be viewed as a species as every other').[12] In the process he raises difficulties for all versions of the realist-essentialist view, as well as for most attempts to classify species on the basis of perceived similarity, evolutionary continuity, individual persistence over time, functional interdependence of parts, and the like. On the other hand, this need not drive us to reject that 'peculiar feature of our concept of an organism' which regards these and other individuating properties as somehow intrinsic to the very idea that x belongs to species y. 'I mention these issues', Sober writes,

> to convey a feeling for just how intricate our ordinary concept of an enduring organism is. It is serviceable enough in most everyday contexts. However, I do not doubt that situations can be envisaged in which it is unclear how the concept should be deployed. The lesson, of course, is that we should not reject the concept. Rather, we must recognize that our concepts are not logically perfect. They, like organisms themselves, get along reasonably well in their normal habitats but may be seriously ill-suited to coping with unusual circumstances.[13]

Sober discounts what he calls the thesis of *trivial semantic conventionalism*, that which holds – in post-structuralist fashion or after the manner of Foucault in some well-known passages from *The Order of Things* – that 'it is up to us what meanings we assign to the terminology we use', so that 'there is nothing inherent in rocks that forces us to call them by the word "rock", rather than by the word "mush" '.[14] Such conventionalism (also implicit in, or derivable from, some of Quine's more extreme formulations of ontological relativity) is 'a universal and hence trivial thesis about how we pair words with meanings'.[15] It must therefore be distinguished from those other (non-trivial) forms of anti-essentialist argument which pose a real challenge to received – commonsense or specialized – ideas about species individuation. Still, 'it is worth remembering that we feel quite reasonable in using the concept of an organism even though philosophers have yet to produce a fully adequate theory of the logic of that concept'. And in Sober's view there is a strong connection – if one that has yet to be worked out in detail – between the sorts of concept by means of which we individuate organisms and those by which we assign them to species on the basis of enduring (ontogenetic and phylogenetic) attributes. At any rate there seems little prospect of success for any theory that adopts an all-out conventionalist or anti-realist approach.

So Empson has some good evidence on his side – evidence from the natural sciences as well as from recent debates in philosophy of science and language – when he takes it that 'dog' is a species-concept whose extension (members of the canine kind) can be picked out with a fair degree of assurance. In *Seven Types of Ambiguity* he had broached this issue by way of some no doubt Russell-inspired reflections on the kind of analysis that might be required in order to fully disambiguate a sentence like 'The brown cat sat on the red mat'. (Incidentally: that this sentence has to do with *cats* and not *dogs*, yet obviously raises the same sorts of issue, is one clear sign that species-concepts and natural-kind terms are sufficiently well understood for most practical purposes.) The passage is distinctly parodic in tone but is worth citing here for the light it throws on Empson's attitude to the then emergent programme of mainstream analytic philosophy. Thus:

'The brown cat sat on the red mat' may be split up into a series: 'This is a statement about a cat. The cat the statement is about is brown', and so forth. Each such simple statement may be translated into a complicated statement which employs other terms; thus you are now faced with the task of explaining what a 'cat' is;

and each such complexity may again be analysed into a simple series; thus each of the things that go to make up a 'cat' will stand in some spatial relation to the 'mat'. 'Explanation', by choice of terms, may be carried in any direction the explainer wishes; thus to translate and analyse the notion of 'sat' might involve a course in anatomy; the notion of 'on' a theory of gravitation.[16]

And for 'cat', needless to say, a theory of species-membership or natural-kind denomination which served to specify exactly what range of creatures – defined in terms of their appearance, attributes, genetic constitution, or whatever – should properly be taken as falling under that concept. Empson offers the example only to make it clear that he is not very interested in this sort of analysis. It would, he continues, be largely irrelevant 'not only to my object in this essay but to the context implied by the statement, the person to whom it seems to be addressed, and the purpose for which it seems to be addressed to him.'[17] In short, 'ambiguity' for Empson is not so much a matter of referential fuzziness (or problems in delimiting the extension of terms) as of the multiple meanings – the intensional complexities – that are brought into play when those terms are deployed in a communicative context.

Of course this approach must take it for granted that we (speakers and interpreters, poets and readers) start out with a good working grasp of language in its extensional or referential aspect, and are thus able to perform those other, more complex kinds of operation. So when we come across the word 'cat', Empson commonsensically remarks, we are not forthwith suspended in some state of agonized indecision as between (for instance) Persian cats, Siamese cats, black cats, white cats, tabby cats, or whatever. Quite simply, we resolve the matter by adopting whatever degree of specificity seems to suit the context, and – beyond that – relying on our best intuitions with regard to some natural-kind concept or paradigm class of such creatures. Here also his assumption is strongly borne out by recent work among cognitive psychologists and proponents of a 'naturalized epistemology' that takes a lead from Quine's well-known use of that slogan while resisting his more sceptical (onto-logical-relativist) conclusions.[18] For on this latter account, as Hilary Kornblith puts it, '[t]he classification of animals into dogs and cats is merely a product of human beings'. And again: '[i]f someone wished to group together under a single term all those animals which were born on a Tuesday, this grouping would, admittedly, be strange. It would not, however, be either accurate or inaccurate. Claims made using this classification could be accurate or inaccurate, but the classification itself would be neither'.[19] This opens the way, not only to Quinean talk of ontological relativity, but also to the exotic thought-experiment which Foucault deploys, in a famous passage of *The Order of Things*, as proof of the 'conventional' or 'arbitrary' character of natural-kind concepts.[20]

However, as Kornblith observes, such scepticism cannot be justified by the mere fact that various samples or exemplars of some candidate kind may appear dissimilar in numerous respects. After all,

> two samples of gold need not be of the same size, shape or weight; they will inevitably differ in a great many of their properties. And when we leave the world of chemistry for, say, biology, the differences among members of the same natural kind become even more salient: a Russian wolfhound and a Mexican hairless are both dogs, yet their properties are dramatically different; even two Russian wolfhounds will differ from one another in a wide variety of ways. Nevertheless, it is clear that members of the same natural kind must be alike in important respects. The problem here is to say in just what respects members of the same natural kind must be alike.[21]

Kornblith's proposed answer to this problem is in terms of those 'homeostatic property-clusters' – subatomic configurations, molecular structures, chemical bonds, DNA genotypes, and so forth – whose presence or absence enables us to draw the distinction between real and nominal kinds. Thus for instance, '[w]hen we come to understand how chemical bonds are formed, we see why H_2O is a possible molecule, but HO_2 is not. The clustering of observable properties is a by-product of the stable configurations which are possible at the unobservable level'.[22] And the same applies to those other natural-kind terms (like 'dog' and 'cat') whose individuating properties are less well-defined since they encompass a far greater range of attributes, from the microstructural (genetic) level to that of appearance, behavioural traits, reproductive habits, or whatever. Nevertheless our capacity to use such terms – to distinguish (say) a cat-like mewling Pekinese from a hairy dog-like growling tom – cannot be accounted for except on the basis of natural-kind identity. Genetically speaking there is further evidence for species

individuation in the well-known phenomenon of second-genera-
tion hybrid sterility, that is, the fact that cross-species offspring are
unable to reproduce.

Of course this is not to argue that such items of relatively
specialized knowledge are somehow prerequisite to our everyday
capacity for picking out dogs, cats, and the rest. Nor, for that
matter, do we need to know the laws of chemical bonding and the
molecular structure of H_2O in order to identify samples of water, or
the fact that gold has the atomic number 79 in order to use the term
'gold' with a good chance of referring to the right sort of stuff.
Rather we rely on what Putnam calls the 'linguistic division of
labour', that is, the knowledge that there have been – and are –
people with acquired expertise in the relevant fields (biologists,
chemists, zoologists and others) who could offer advice if needed.[23]
Putnam first introduced this idea with a view to resolving one
obvious problem with the causal theory of reference. For in the
standard (Kripkean) version it failed to explain how speakers were
able successfully to deploy a vast range of referring expressions
despite having picked them up at second hand, by merely hearsay
acquaintance, or without any possible linkage back to the inaugural
moment of naming *via* the postulated causal chain of transmission.[24] But this also suggests – as some commentators have
argued – that the theory should be viewed as an additional resource
rather than a full-scale alternative or rival to the other (intensional-
descriptivist) account of how terms acquire sense and reference.[25]
For if reference depends, to some extent at least, on the linguistic
'division of labour', then it follows that speakers and interpreters
outside the causal chain will be able to use those terms correctly –
or to designate the right sorts of object – only in virtue of some prior
acquaintance with their senses, meanings, or criteria for proper
application. Thus the causal theory cannot do without a descriptivist component, any more than descriptivism – in its purebred version – can overcome the problems and counter-intuitive
consequences that Kripke so shrewdly points out.

By this time the reader may well be wondering just why I have
taken such a long way around to the promised discussion of
Empson's chapter on 'The English Dog'. One reason, as already
mentioned, is that Empson here takes certain things for granted as
regards the status of referring expressions and – more especially –
the existence of natural kinds (like cats and dogs) to which those
expressions may be taken to apply without too many Russell-type

analytic scruples. Another is the way that he, like Putnam, bases his approach on a certain idea of the linguistic 'division of labour'. In Empson's case this involves a working assumption that even if terms are to some degree 'fixed' through the reference-back to some paradigm instance, still there is a great deal more to the process through which they acquire a socialized currency of meaning. In fact he is less concerned with *literal* uses of the word 'dog' – i.e., those that pick out members of the canine kind, whether through an act of 'rigid designation' in the Kripkean manner or some cluster of descriptive attributes – than with the rich variety of metaphoric, colloquial, or slang associations that accrued to the word at different periods and in different socio-cultural contexts of usage. All the same this process could scarcely have got started – or have made any sense in linguistic or communicative terms – were it not for the primary meaning of 'dog' as defined through its reference to *just that kind* of familiar (despised or affectionately regarded) creature.

Thus on the one hand 'sense determines reference' in so far as doggishness is here treated as a complex of attributes, properties, or traits that are taken to characterize paradigm members of the kind and can hence be extended – by analogy – to human beings thought of as exhibiting similar (good or bad) qualities. But on the other hand 'reference determines sense' in so far as the sorts of quality involved are considered intrinsically a part of canine nature and thus as making it somehow *natural* – appropriate, just, or fitting – that the word 'dog' should on occasion be used in the context of human address. It may not always be clear, as Empson remarks, 'how far this feeling would apply to actual dogs', since once the process begins it is likely to produce a whole range of complex attitudinal shifts which may go beyond any credible naturalistic appeal to what dogs are *really* (biologically or instinctually) like. And of course there is a sense in which Empson's approach *via* the semantics of complex words is one that inherently tends to privilege the intensional (meaning-based) mode of linguistic interpretation rather than the kind of extensional analysis favoured by causal theorists such as Kripke and – with significant qualifications – Hilary Putnam. Still it is quite evident, Empson thinks, that 'very mixed feelings are there to be drawn upon'. And since 'dog' became a focus for just those feelings at a crucial stage in their development there would seem good reason to suppose that dogs – 'actual' dogs of the living, breathing, eating, copulating, natural

kind – must always have played some role in the process as a more
or less stable (if scarcely 'fixed') point of reference.

I must now give some idea of how this all works out in Empson's
chapter as a matter of detailed logico-semantic and socio-cultural
interpretation. (That he manages to bring these two dimensions
together – and in a way that finds no parallel in current philosophical semantics – is I think the most impressive achievement of *Complex Words*.) The first point to note about 'dog', Empson says, is the
drastic change of mood or evaluative tone that the word underwent
between the Renaissance and Restoration periods. Thus:

> Before the Restoration the dog of metaphor, by and large, is
> snarling, a sycophant, an underdog, loose in sex and attracted by
> filth, cruel if it dare; 'love me, love my dog' means 'love the
> meanest thing about me'. There is the biblical dog, a pariah, living
> on crumbs and Jezebel as they drop ('a dog's chance' – he is
> dependent on human society and yet friendless in it); also the
> dog-faced Thersites of Homer, a mean and envious mocker
> (staying in the manger, barking at the moon). Shylock is eminent-
> ly a dog of this sort and often called so; a man so placed can
> hardly be expected not to pervert justice, though this is a warning
> for you, not an excuse for him.[26]

It is the same cluster of anti-canine – and also deeply misanthropic
– sentiments that Empson discovers in Shakespeare's *Timon of
Athens*, the subject of his next chapter 'Timon's Dog'.[27] The metaphor is associated chiefly with Apemantus, a cynical and malcontent character whose low view of human nature – rapacious,
opportunist, treacherous, and so forth – comes across in the various
pejorative senses of 'dog' that characterize his own usage and the
usage of others speaking about him. Thus 'Apemantus is continually called "dog" (in a play clearly meant to be taken as symbolic)
with the sense "snarling and envious critic", also as "ill-condi-
tioned" – he will "famish, a dog's death" ' (*Complex Words*, p. 178).
And of course Timon himself comes around to this cynical way of
thinking when he loses his wealth – having lavished it upon his
servile and cringeing (dog-like) flatterers – and then finds himself
spurned as a dog by those same erstwhile boon companions.

However, as Empson points out, there is more going in the
complex semantics of 'dog' than is fully covered by this set of
rock-bottom cynical equations that connote servility, craven oppor-

tunism, time-serving flattery, and the like. For Timon remains oddly impressive – a genuine tragic hero – despite his near-farcical change of character from nobility of soul to an attitude of boundless cynical contempt for humanity at large. Empson cites a passage from Erasmus (*The Praise of Folly*) that asks: 'What is more fawning than a spaniel? Yet what more faithful to his master?'. And he finds the double sentiment still at work in *Timon of Athens*, though now under strain – almost to breaking-point – from the outright collision of attitudes which (as many critics have felt) makes this such a deeply problematical play. Thus when Timon and Apemantus meet before the cave and exchange insults 'each has strong grounds for priding himself on his version of self-contempt and despising the other's'. Like 'honest' in *Othello* or 'sense' in *Measure for Measure* the word 'dog' is here the focus of conflicting valuations which reflect a much larger cultural clash, roughly speaking that between Renaissance humanism and an emergent puritanical ethos which stresses Original Sin and the brutishness of unregenerate human nature. As Empson puts it:

> Shakespeare is both presenting and refusing a set of feelings about *dog* as metaphor, making it in effect a term of praise, which were already in view and became a stock sentiment after the Restoration. It is a popular but tactfully suppressed grievance that Shakespeare did not love dogs as he should, and I think the topic is a large one; when you call a man a dog with obscure praise, or treat a dog as half-human, you do not much believe in the Fall of Man, you assume a rationalist view of man as the most triumphant of the animals (*Complex Words*, p. 176).

Clearly there is a strong normative dimension here, an appeal to certain secular-humanist values and beliefs carried by a range of associated keywords such as 'sense', 'honest' and 'dog'. These in turn provide a background of enabling presuppositions which Empson himself both strongly endorses – as a matter of personal commitment – and considers typical of the way that such words take effect, even in cases like *Timon* where their meaning may be twisted almost beyond recognition through the conflict with opposing (cynical, religious or anti-humanist) doctrines.

By the Restoration period, with puritan values in retreat, the secularizing impulse emerged once again – reviving the spirit of early humanists like Erasmus – though often in a negative or reactive form (at its most extreme in the poet Rochester) which preserved something of the same deep-laid ambivalence. Thus '"dog", it is absurd but half-true to say, became to the eighteenth-century sceptic what God had been to his ancestors, the last security behind human values' (*Complex Words*, p. 168). On the one hand, it could carry doctrines (or semantic equations) of the type: 'Men are no more than animals. No man by any effort can escape the charmed circle of self-interest' (ibid.). But on the other, there persisted the alternative, more hopeful or generous idea of what humans might share with the lower animals and yet come off rather well from the comparison. In its rock-bottom form: 'yet if the worst is the dog, humanity is still tolerable'. And from this point the idea could be built up into something like a naturalized human-ist ethic, a kind of self-implicating irony that also – by virtue of its shared possession – entailed a fair measure of other-regarding sentiment. Thus Empson acknowledges a debt to Wilson Knight for having first pointed out the importance of 'dog' as a key metaphor in Shakespeare's play. But he is also sure that Knight 'oversim-plifies' by 'making Apemantus a mere symbol of Nasty Cynicism and Hate' (*Complex Words*, p. 176). For despite all the clearly pejor-ative uses – here as elsewhere in Shakespeare – there are also some instances which do seem to carry that 'obscure hint of praise' (about animal and human nature) which was firmly suppressed by exponents of the Christian-orthodox view.

What is impressive about Empson's reading is its desire to do justice to a play which most critics have found either downright repellent or so utterly bleak and misanthropic that it fails both as tragedy (since we just cannot sympathize with Timon) and as satire (since it lacks any countervailing sense of positive ethical or social values). At very least, Empson argues, even though 'nobody pre-tends that *Timon* is a good play', still, 'given the Malcontent theme this is the cleverest treatment of it ever written' (p. 180). But really he is pitching the claim much higher since the kind of 'cleverness' (or linguistic resourcefulness) that can find something more than rock-bottom cynicism in a word like Shakespeare's 'dog' is also the kind that enables speakers and interpreters to communicate across otherwise sizable gulfs of background presupposition and belief. Of course Empson is not denying that this process can sometimes break down and produce all manner of mutually intensified baffle-ment, mistrust, and downright contempt. Thus 'in the end the only effect of this idea [i.e., the cluster of dog-related feelings] is that

Timon can go off into neurotic insanity, and Apemantus is forced for contrast into a reasonable view of life in which he can cut no figure' (p. 183). Still the only way that these ironies can achieve so powerful a dramatic effect is through our grasping the background of normative assumptions against which they then stand out – in stark and tragic relief – as one extreme of the range of sentiments that might be carried by the word.

Thus the main point about such words, Empson suggests, is that they typically evoke a certain 'humour of mutuality', a sense that we are all (the cynic included) somehow in the same boat, or entitled to the same sorts of generous allowance for the problems and perplexities of human life. This is what makes it difficult – though on occasion a source of powerful dramatic effect – to wrench them away from that normative (secular-humanist) background and push them toward the perversely ascetic end of the scale, that is to say, the puritanical 'pleasure in self-contempt' that typifies Timon's usage. 'Men treat Timon and Apemantus as dogs; Timon and Apemantus notice at last that they cannot escape being men, so that there is some logical puzzle for them in railing against mankind; let them then praise dogs, among whom they are metaphorically included as cynics, and in rebuking man they may half-praise themselves' (*Complex Words*, p. 179). But the logical puzzle remains; they (the *dramatis personae*) may be too little aware of it, thus cutting themselves off from any claim to shared humanity, while we (the audience-members or readers) are better placed to understand how they might have been driven to this crazy extreme.

So it is, on Empson's reading, that '*dog* holds the key positions at both ends of a scale; furthermore it gives a hint of escape from the conflict that made the scale important, because there is a hint of liking for dogs in general' (p. 177). At any rate the word crops up so often in the play – and in contexts that imply such a range of diverse (sometimes contradictory) meanings – that it must have registered, however obscurely, as a test-case for the various attitudes concerned. Thus on the one hand 'dog' could be used of human beings in a strongly pejorative or contemptuous sense, taking dogs to be among the lowest of animal creatures on account of their servile, cringeing nature, their attraction to disgusting smells, their repulsive eating-habits, promiscuous sexuality, and so forth. The two stock sentiments to be drawn upon here were the Christian-orthodox view (man as a creature born in sin and always prone to revert to his base animal nature unless firmly restrained) and the ancient – etymological – link that equated 'dog' with the behaviour of a *cynic* like Diogenes (urinating and defecating in public, masturbating openly in the market-place, setting out to undermine all notions of human dignity and spiritual worth). It is these sentiments that Shakespeare exploits in the most famous of his dog-metaphors from *Antony and Cleopatra*. Thus 'Shakespeare, as has often been pointed out, had a sort of fixation of disgust between the words "spaniel" and "candy"; the dogs appear always slobbering and melting their sweets, and the cynic imagined as *dog* always at bottom wants to get on in the world he despises' (*Complex Words*, p. 165). But on the other hand, the word could not easily be kept from suggesting other, more praiseworthy attributes, among them – to pick out just a few – companionship, trust, tail-wagging affection, perseverence ('doggedness'), truth to its natural instincts, and what Empson calls 'a cheerful stoicism based on independence and indifference to dignity'. Of course these stand in direct, one-for-one opposition to the range of pejorative senses and implications listed above. And yet, Empson thinks, it is precisely by virtue of this deep-laid ambivalence – its readiness to flip over from one to the other set of meanings – that the word took on such a power to evoke whole backgrounds of conflicting ('pro-dog' and 'anti-dog') sentiment. Thus '[v]arious elements which had given the dog strength as a symbol of evil could still give it strength as a term of praise' (p. 167).

My point in all this is that Empson's ruminations on the name and nature of 'dog' have a lot more to tell us about language, meaning and interpretation than might be gathered from their often rather casual and deliberately 'blow-the-gaff' style. This latter is indeed a phrase that he uses (with apologies in a footnote) to suggest one of the word's more prominent feelings, the idea that 'dog' – as applied to fellow-humans – might convey not an insult but a kind of rock-bottom though obscurely generous praise, a hint that dogs are after all not so bad and that canine nature is a decent (if unexalted) metaphor for the human condition in general. Thus '[t]here is enough "blow-the- gaff" feeling about calling a man a dog to give him a fundamental sincerity (as if by reflection from the speaker); he does not hide the truth about himself and thereby shows the truth about us all'. (Empson's footnote explains that he is using the phrase – for want of anything better – to suggest 'something like "give the low-down", with the extra idea of puncturing something inflated'; *Complex Words*, p. 166n.) What this example brings out rather nicely is the normative dimension of Empsonian semantics, that is, the idea that certain human attributes – mutuality, trust, a degree of self-interest but also (not incompatible with this) a fair measure of other-regarding sentiment – are there in the background and ready to be drawn upon when interpreting other people's language and beliefs.

I have suggested already that Empson's approach finds a parallel in certain quarters of recent philosophical debate, notably in the kind of naturalized epistemology (and the concomitant 'Principle of Humanity') proposed by – among others – David Papineau.[28] That is to say, it works on the principle that utterer's meaning can best be construed with reference to *both* a truth-theoretic semantics – yielding propositions which can then figure as contents or objects of belief – *and* a causal-explanatory account of how those beliefs most likely came about, in response to what needs, pressures of circumstance, formative cultural conditions, and so forth. The great virtue of Empson's approach – especially in 'The English Dog' – is to offer a detailed and convincing illustration of the way that these components fit together. For the chief point about 'dog' is that the word involves such a complex range of beliefs having to do not only with members of the canine kind (their nature, behaviour, species-identity, characteristic vices or virtues) but also with humanity considered in relation to the two opposed 'scales' – religious and secular – which between them marked out its range of possible implications. Thus even in *Timon* 'the dog rides the conflict; on the one hand a low creature which distributes contempt, on the other it is good-natured because self-satisfied about its direct pleasures (what else has Timon felt in generosity and glory?)' (*Complex Words*, p. 183).

CANINE VARIATIONS: EMPSON AND SCHIFFER

One likely objection to this whole line of argument is that Empson's cannot be a general theory of semantic interpretation – a theory that should work in principle for all sorts of utterance – since it clearly has an inbuilt elective bias toward the secular-humanist end of the scale. That is, he tends to favour words (such as 'sense', 'honest' and 'dog') whose meaning may sometimes be pushed in an opposite direction but which still carry something of that normative appeal to a background of shared values and beliefs. This preference is obvious enough in his two chapters on 'dog' where the dominant Shakespearean set of equations – 'slobbering', 'dirty', 'cynical', 'fawning', 'hypocritical', 'slave to base instincts' – is treated as very much a special case or an instance of the deep-laid (even psychopathological) conflict induced by the clash between opposed ideologies of secular and puritan-religious sentiment. Thus in *King Lear* also 'they [dogs] chiefly appear as snobs, cruel to the unfortunate, also either as flatterers or as habitually flattered. . . . The Fool has a phrase "dog in madness" which apparently appeals to hydrophobia, in a determined effort to make metaphorical dogs join the variety of lunatics already represented in the play' (*Complex Words*, pp. 165–6). All the same – and here Empson's normative bias comes out – 'this is no part of the ordinary symbolism, which makes dogs typically, even if offensively, sane' (p. 166).

In short, there is still that strongly marked normative background of tacit admiration for dogs – and for human beings in their more doglike or shared animal nature – which allows us to register this cynical mood as a definite departure from the norm, a perverse effort to exclude or suppress the more natural range of senses in the word. In *King Lear*, Empson thinks, 'Shakespeare had thus reached the height of his anti-dog sentiment just before the mysterious pro-dog gestures which are part of the cynicism of *Timon*' (p. 166). And again this cynicism achieves its most powerful (tragic though at times near-farcical) effects through our retaining some awareness of the other range of uses – 'trusting', 'affectionate', 'true to its instincts' – which are still implicit in the word. For this is what enables 'dog' to convey an alternative, more hopeful or generous idea of human nature and the prospects for mutual understanding, a theory with substantive social and ethical implications as well as a theory of 'complex words' and their logico-semantic structures. Which is also to say – and here Empson agrees with Papineau – that interpretation-theory *cannot do without* some implied normative dimension, some 'Principle of Humanity', broadly conceived, that gives content to the otherwise arid formalisms of a purebred analytical account.[29]

'Sense' is another good example here since its meanings cover such a massive range, from empiricist or phenomenalist doctrines at the one end ('sense' = 'senory impressions' or 'sense-data') to Wordsworth's mystical-pantheist 'language of the sense'.[30] In the latter case it is through his use of paradoxical 'Type IV' equations that Wordsworth – according to Empson – achieves his effect, equations (that is) whose 'claim is false' because they 'do not really erect a third concept as they pretend to' (*Complex Words*, p. 305). And again: 'the whole poetical and philosophical effect comes from a violent junction of sensedata to the divine imagination given by love, and the middle term is cut out' (p. 296). Yet there must still be at least some tacit appeal to that middle-ground range of meanings – 'good sense', 'commonsense', 'sound judgement' – in the absence of which these paradoxes would fail to register, or the word just disintegrate for lack of any normative (humanly intelligible) structure. At the opposite extreme, in Shakespeare's *Measure for Measure*, 'sense' gets pushed toward a rock-bottom cynical interpretation (like Timon's contemptuous use of 'dog') where its meaning comes out as a warped and hideous inversion of the puritan creed. Thus Angelo, describing Isabella: 'She speaks and 'tis/Such sense, that my sense breeds with it. Fare you well.' The first use here, Empson thinks, is something like 'wise or reasonable meaning', but then ' "sensuality" . . . pokes itself forward and is gratified by the second use of the word as a pun' (p. 274). And as Angelo (at this point 'genuinely suprised by his desires') is thereafter driven to the furthest extremes of corrupt and self-hating sensuality, so the keyword 'sense' undergoes all manner of perverse or deviant semantic twists.

The play makes an interesting test-case for Empson's method since it seems to force a series of equations into the word which entirely lose contact with its 'normal' or 'middle-ground' interpretative range. There were three chief meanings current at this time, Empson thinks: (1) 'sensuality', (2) 'sensibility', and (3) 'sensibleness', though this latter was as yet not fully developed and entered most often by loose analogy with the ideas of 'a truth-giving feeling' or 'a reasonable meaning'. What happens in Angelo's soliloquies – and here I offer a reductive paraphrase of Empson's far subtler and more detailed analysis – is that (1) takes on an intense but wholly pejorative range of feelings which assimilate (2), thus preventing 'sensibility' from exerting any humane influence, and (3) is effectively cut out altogether as Angelo descends into a state of near-insane sexual fixation and obsessional neurosis. 'Clearly', Empson writes,

the equations between these three could carry very relevant ironies, though the effect is not so much a covert assertion as something best translated into questions. Are puritans hard? (Is not-one not-two)? Are they liable to have crazy outbreaks? (Is not-one not-three?) Is mere justice enough? (Is three two?) To be sure these questions look very unlike the flat false identity of one idea with another; but I think the state of the word then made them easier to impose. It seems to have been neither analysed nor taken as simple; it points directly into the situation where it is used, implying a background of ideas which can be applied to the situation, but somehow as if the word itself did not name them; it is a shorthand term, rather than a solid word in which two of the meanings can be equated. And yet, as the play works itself out, there is a sort of examination of the word as a whole, of all that it covers in the cases where it can be used rightly; or rather an examination of sanity itself, which is seen crumbling and dissolving in the soliloquies of Angelo (*Complex Words*, p. 270).

I have quoted this passage at length because it brings out everything most distinctive, original, and (I would argue) philosophically pertinent in Empson's approach to the semantics of complex words. For one thing it shows – what Davidson misses – the extent to which our handling of language may involve great refinements and subtleties of logico-semantic grasp, tacit understandings that cannot be explained by recourse to a minimalist ('passing') theory of utterer's meaning.[31] Then again, it draws the distinction between 'flat false identity' of the kind involved, say, in Orwell's mind-wrenching paradoxes – such as 'War is Peace' – and those other, more suggestive (or subtly insinuating) forms of aberrant equation whose illogicality tends to escape notice.[32] This is because they inhabit that highly ambiguous zone between foreground and background, 'chief' and 'head' meaning, or structures that are felt to be 'there' in the word – in its compacted propositional sense – and implications that are felt more vaguely to evoke a whole context of associated (sometimes conflictual) attitudes, values, and beliefs. But there is also the claim – strongly conveyed in Empson's last sentence – that even the most puzzling cases of this sort leave open the appeal to some larger dimension of shared linguistic understanding against which they are viewed as singular departures from the norm and hence as instances of irrational belief, pathological obsession, 'dramatic irony', or whatever.

So it is that finally '[the play] moves over, as the key-word does, from a consideration of "sensuality" to a consideration of "sanity", and then the action is forced round to a happy ending' (*Complex*

Words, p. 287). This ending is 'forced' in the sense that it still leaves some crucial problems unresolved, among them a range of much-discussed thematic problems (about justice, sexuality, puritanism, 'human' *versus* 'divine' law, etc.) and also – corresponding to these – problems of a logico-semantic character having to do with 'sense' and its liability to just such deviant or aberrant uses. Certainly the paradoxes are here intensified to a point where the middle-ground meanings are almost squeezed out and where analysis finds itself very nearly played off the field. Thus Empson betakes himself to a style of paraphrase more in the *Seven Types* manner: '[t]he subtle confusion of the word is used for a mood of fretting and exhausted casuistry; the corruption of the best makes it the worst; charity is good, but has strange and shameful roots; the idea of a lawsuit about such matters is itself shameful, and indeed more corrupt than the natural evil' (p. 273). But he has also shown – and I think to highly convincing effect – that the process of arriving at these finally rather baffled and paradoxical conclusions is one that has to go through all the complex operations that characterize our other (more normal or everyday) modes of linguistic understanding. Nothing could be further from Davidson's idea that 'wit, luck, and wisdom' are all it takes to figure out a suitable passing theory for any odd or contextually deviant utterance that might happen to turn up.[33]

What I have been talking about mainly up to now is the *intensional* aspect of complex words, that is to say, their propositional content as given through a logico-semantic analysis on the lines that Empson suggests. But he is also very clear that we cannot make sense of such words unless there is some way of relating that content to the attitudes, beliefs or *intentions* which speakers express through particular forms of utterance in particular contexts of usage. Such – we may recall – was the aim of Schiffer's strenuous but forlorn quest: an intention-based semantics that would make it possible to give an adequate truth-theoretic account of both sentence-meaning (propositionally construed) and utterer's meaning (construed in relation to sentences as objects or contents of belief).[34] If this project has failed, as Schiffer now thinks, then so must any other attempt to link up the various component parts – reference and sense, extension and intension, belief-contents and propositional attitudes, sentence-meaning and utterer's meaning – through a theory somehow immune to such criticism. Quite simply, there is no part of that original programme left standing and no possible

alternative theory that would not require at least *some* items on the
above list to have survived the sceptical rigours of Schiffer's analysis. Thus extensionalist accounts must ultimately fail through the
lack of any sure, unambiguous criterion for picking out referents,
objects, or classes of object on the basis of reliable (e.g. natural-
kind) identity. Intensionalist theories are no better placed since
they appeal to semantical (truth-theoretic) notions of analyticity,
propositional structure, or meaning-invariance across differing
contexts of utterance which likewise prove inadequate to the task.
And if this is the case then there cannot be much hope for any form
of intention-based semantics that would treat such (putative) contents of belief as the basis for a theory of propositional attitudes. For
we then have a threefold problem as to (1) how objects are identified, (2) how propositions or belief-contents are individuated in
respect of those objects, and (3) how such contents if indeed they
exist – *concesso non dato* – could ever be equated with any specific
intention on the speaker's part. Thus in the case of 'Tanya believes
that Gustav is a dog', there is just no way – on Schiffer's account –
that the sentence could be unpacked with anything like the required degree of conceptual and truth-preserving rigour.

Still one may respond that such sentences are very often used,
and such beliefs entertained, by speakers and interpreters who
might reasonably claim to be in no doubt on any of the above
points. That is, we should need very strong evidence before con-
cluding that Tanya was mistaken about dogs in general, about this
dog (Gustav) in particular, about the content of her own belief
regarding Gustav, or again, about the sense of the referring expression ('Gustav is a dog') which conveyed her intent to say just
that about the creature concerned. Of course this is not to deny that
we *might* on occasion possess such evidence and decide that she
was indeed mistaken. But then we should not only be interpreting
her words – attributing beliefs on a truth-theoretic and intentional
basis – but also looking around for some causal-explanatory factor
that could best make sense of an otherwise deviant or off-the-point
utterance.[35] This, to repeat, is where 'humanity' comes in, as distinct from that all-purpose Davidsonian principle of 'charity' which
simply counts her right, come what may, for want of any better
(more adequate) alternative theory. It would then be a matter of
adducing grounds – humanly intelligible grounds – that could
justify our counting her wrong in this particular instance, that is,
with respect to Gustav, dogs, natural kinds, class-membership

criteria, the logical grammar of referring expressions, self-ascribed belief-states, and so forth. Or the case might require some reckoning with the sorts of special circumstance – perceptual distortion, unusual lighting conditions, misinformation, habitual errors of usage, and so on – which Papineau brings up in support of his causal-explanatory theory. Anyway the point can be made more simply: that figuring out the sense of what people say is a process at once more humanly complex and less ultimately mysterious (or downright inexplicable) than Schiffer contrives to suggest.

There is, to my knowledge, no discussion in the recent philosophical literature that approaches Empson's in its capacity to bridge the widening gulf – described with such plaintive eloquence in Schiffer's book – between natural-language ('commonsense') understanding on the one hand and truth-theoretic formal semantics on the other. I venture this claim with some trepidation, given that Empson shows scant regard for any 'theory' or 'philosophy' that gets out of touch with the practical-interpretative business in hand. Thus 'the whole notion of the scientist [or, as it might be, student of philosophical semantics] viewing language from outside and above is a fallacy; we would have no hope of dealing with the subject if we had not a rich obscure practical knowledge from which to extract the theoretical' (*Complex Words*, p. 438). But it is here precisely that Empson's approach diverges from the analytic mainstream and thereby avoids the sorts of dead-end encountered in their different ways by Schiffer and Davidson. His point is that theorizing can indeed be useful – else why so much of it in *Complex Words*? – but only in so far as it helps to clarify the sources of that 'rich obscure practical knowledge' that must be at work if speakers and interpreters (or poets and readers) are to communicate at all.

This is why I would suggest that philosophers do have something to learn from literary theory, despite their (often well-founded) suspicion that much of what currently goes under that name is philosophically naive, ill-informed, and confused. For it is among Empson's most firmly-held convictions that the kinds of complexity to be found in literary langage differ in degree only and not in kind from those that characterize our 'ordinary' (everyday-communicative) modes of talk. Hence his quarrel with the New Critics over their idea of poetry as belonging to a realm apart, an autonomous domain of metaphor, paradox, irony, or other such privileged rhetorical tropes which placed it (supposedly) beyond reach of the plain-prose rational intellect.[36] This he sees as nothing more than an obscurantist drive to protect their own status as custodians of the sacred text while discouraging any non-initiate reader who might ask naively what the author meant to say, how far he or she was successful in the attempt, and whether or not the intended meaning – the structure of logico-semantic implication – was able to support the truth-claims advanced in however 'paradoxical' a form.[37] What we have here, in short, is a point for point reversal not only of the New Critics' doctrine but also of that minimalist-semantic line in philosophy of language that despairs of discovering any viable relation between truth, meaning, and utterer's intention construed in belief-based (propositional) terms. Moreover it is Empson's great virtue as a close-reader of texts that he is able to combine, like few others, a keen responsiveness to verbal detail with an equally keen analytical intelligence that keeps these questions constantly in view.

There is a nice example of how this all works out in his thoughts about Dr Johnson on the subject of dogs. In his more 'official' or moralizing statements Johnson took the Christian-orthodox view that human nature was inherently bad – the result of Original Sin – and hence that its lowest (animal) proclivities must be held firmly in check. This made him ultra-sensitive to the cluster of feelings that had developed around 'dog', the secular-humanist idea of man as 'most triumphant of the animals' and of dogs as embodying the best – rather than the worst – in the range of instincts, needs and desires that humankind shared with the animals. Thus Johnson famously denounced Fielding's *Tom Jones* as a novel that tended to encourage such dangerous ideas through its treatment of Tom's various (mainly sexual) peccadilloes as unimportant in comparison with his natural goodness and generosity of spirit. The effect of this doctrine, Boswell reports him as saying, was to 'resolve virtue into good affections, in contradistinction to moral obligation and a sense of duty'. And again: Fielding was 'the inventor of the cant phrase, goodness of heart, which is every day used as a substitute for probity, and means little more than the virtue of a horse or dog'.[38] What Empson finds interesting about this last passage is that 'Johnson here has to use a synonym for *honesty* and put *horse* beside *dog* to keep the slang senses from rising at him' (*Complex Words*, p. 173). For 'honest' is another of those words – as Empson shows elsewhere in the book – which by then had acquired a whole range of complex semantic implications. These extended from the malign, cynical usage that Iago deploys against Othello ('honest = 'ready to acknowledge and exploit the worst elements in oneself and others') to the far more generous and tolerant idea: 'honest = true to oneself, aware of one's own and of others' faults, hence tending to take an unillusioned but reasonably sanguine view of human nature'). It is this latter set of meanings that Johnson finds so abhorrent and which he therefore needs to exclude by *not* using 'honesty' – as might have been expected in context – but instead the word 'probity' with its stronger appeal to rigorous standards of moral rectitude. And with 'dog' likewise there is the sense of a certain strategy at work, a decision – at whatever unconscious or preconscious level – to place the word 'horse' alongside it and thus prevent 'dog' from calling up an unwanted range of (as Johnson thought them) perverse and mischievous sentiments.

What makes the case yet more complicated is the fact that two rather different though related strands of thought had both become attached to the same set of words. They can roughly be distinguished, Empson suggests, as 'rationalist' and 'humanist', with the former tending to lay more stress on moderation, self-discipline and the classical virtues, while the latter conveys an acceptance of the naturalist ethic – 'truth to one's instincts', 'dogged veracity' and the like – which came out sometimes in marked opposition to the rationalist view. Dr Johnson's opting for 'horse' as a counterweight to 'dog' can be felt to pick up on this distinction since it suggests the familiar idea (as in Swift) that horses are eminently sane and 'rational' creatures not prone to the sorts of brutish behaviour that are only to be expected of dogs. Thus there seemed a fair prospect of achieving some middle-ground position – what literary critics, Paul Fussell among them, have termed 'Augustan humanism' – which could reconcile the values of a broadly Christian morality with those of a broadly secular belief in reason as the chief civilizing influence in human affairs.[39]

However this balance was not easy to hold since (as Empson remarks about Johnson) the other set of meanings could always rise up against the effort to pursue such a moderate course. The following passage from *Complex Words* suggests how perilously thin was the line that separated the Augustan concordat from other, more heterodox (or naturalized) varieties of humanist ethic. 'The eighteenth-century rationalist', Empson writes,

> limited very sharply the impulses or shades of feeling he was prepared to foster – not merely Enthusiasm was cut out but the

kind of richness of language that Gray lamented over in a passage
from Shakespeare – and yet could pursue Reason with gusto and
breadth, without emotional skimpiness; there is a savour in his
work which a man such as Herbert Spencer has lost. The feeling
that the dog blows the gaff on human nature somehow attached
itself to the ambition of the thinker to do the same, and this helped
to make him cheerful and goodhumoured. His view of our nature
started out from a solid rock-bottom, a dog-nature, which his
analysis would certainly not break by digging down to it; this
made him feel that the game was safe, and the field small enough
to be knowable (pp. 168–9).

Obviously Empson has a good deal of sympathy with this outlook.
In *Seven Types* he had defended his hard-pressed analytical approach – and shrewdly pre-empted hostile reviewers – by using the
same metaphor, that of 'digging down' as far as one could go
toward the roots of poetic response, but without (as he argued) any
risk of thus destroying the sensitive plant. After all, a good poem
could hardly be damaged by the process, and might just emerge
with its meaning enhanced by the critic's exegetical labours.[40] In his
second book, *Some Versions of Pastoral*, there is a chapter on Milton
which also – more obliquely – puts the case for this approach as
opposed to the kinds of 'appreciative' criticism that would shield
poetry from any such intrusive meddling by the plain-prose intellect.[41] This essay goes by way of the eighteenth-century editors and
critics, among them the scholar Richard Bentley whose rationalist
glosses on *Paradise Lost* have mostly been viewed as an outright
defeat for the analytic method pushed to an absurd extreme.
Empson acknowledges that the method had its failures, and that
Bentley was often out of his depth when confronted with the gran-
deur and pathos, the mind-wrenching tensions and paradoxes of
Milton's poetry. Still he thinks it on balance much more unfortunate – indeed something of a disaster – that the episode has since
gone down as a cautionary tale about the folly of adopting a 'rationalist' approach to Milton or to poetry in general. For the effect has
been to discourage critics from asking the right sorts of question for
fear of coming up (like Bentley) with the wrong sorts of answer.

'In any case', Empson writes, 'it is refreshing to see the irruption
of his firm sense into Milton's world of harsh and hypnotic, superb
and crotchety isolation'.[42] And he then goes on to adduce various
passages in the poem and in Bentley's commentary where the

'method' enabled him to raise questions – grammatical, stylistic, theological, and moral – which simply dropped out of sight for later critics of a more orthodox, less rationalist bent. It was just these questions that Empson took up in his own later book *Milton's God*, a book that has received much the same treatment – and from much the same quarters – as was meted out to Bentley by the guardians of religious and literary-critical decorum.[43] 'That his failure was so crashing was [Empson thinks] a melancholy accident; it was not that his methods were wrong but that the mind of Milton was very puzzling, as his methods showed.' Moreover, it is a topic that urgently needs considering 'because Bentley has been used as a bogey; he scared later English critics into an anxiety to show that they were sympathetic and did not care about the sense'.[44] This is scarcely a charge that could be laid against Empson on the evidence of that chapter in *Complex Words* where he attempts to draw up a list of the senses – or the structures of compacted logico-semantic 'equation' – that are carried by Milton's near-obsessional use of the encompassing keyword 'all'.[45] In the end, as he admits, this attempt fails; the word seems to function more like a 'Freudian symbol' or a 'Wagnerian motif'; and 'so far from being able to chart a structure of related meanings . . . you get an obviously important word for which an Emotive Theory seems about all that you can hold' (*Complex Words*, p. 101). Or perhaps it functions merely as a 'logical connective' that pulls together such a range of senses from so many different (obscurely associated) contexts that the reader is able to connect them only in this 'vague and distant' way. At any rate one can see – thinking back to the chapter in *Some Versions* – why Empson should have come out so firmly in Bentley's defence. For even if the method ultimately fails with a special case like Milton we can still learn more from that failure – more about Milton, about complex words, about the nature of linguistic understanding in general – than could ever be learned through a downright refusal (on religious or aesthetic grounds) to adopt this analytical approach.

Dr Johnson had his own problems with Milton but even more so with that emergent secularizing ethos whose two main elements, according to Empson, 'can be distinguished roughly as rationalist and humanist'. The latter is what Johnson so hated about *Tom Jones*: the belief that human nature is basically good, that instincts (or affections) are the best guide to morals, and hence that there is nothing shameful – indeed some cause for satisfaction – in the idea

of man as sharing those instincts with dogs and other (albeit select)
denizens of the animal world. The former is what finds its way into
Bentley's glosses: a stress on the prosaic virtues of reason, moderation, critical enquiry and disciplined scholarly method. Its effect is
not merely to cut out 'Enthusiasm' – in the religious sense of that
word – but also (as with Bentley) to induce a certain attitude of
principled mistrust toward the 'richness of language' that had
somehow been lost since the period between Shakespeare and Milton. This is not so far from Johnson's feeling – witnessed in his
Preface to Shakespeare and *Lives of the Poets* – that the 'richness' in
question was also a form of near-anarchic verbal licence, a prone-
ness to ill-disciplined profusions of metaphor, ambiguity, 'con-
ceited' wordplay and the like.[46] Behind it there is the sense that
these linguistic aberrations were connected with (or harbingers of)
the breakdown into civil and religious strife that had overtaken
English society in the mid-seventeenth century.[47] Thus Johnson
could accept at least one side of this rational-humanist complex of
sentiments, that which stressed the more prosaic virtues (reason,
self-discipline, a sensible acceptance of human limitations) and
offered a defence against 'enthusiasm' in all its religious and secular forms. '*Tom Jones*, he thought, was wicked because it was an
attempt to make the hearty dog-sentiment into a system of morality; whereas in serious talk you must keep to the rationalist one,
as a sane recognition of the Old Adam' (*Complex Words*, p. 173).

However it is a main point of Empson's argument that 'serious
talk' of the kind that carries doctrines, official creeds, orthodox
statements of belief and so forth is not – or not always – the best
place to look if one wants to discover the truth about human
feelings, meanings and intentions. Thus he finds it revealing 'that
Johnson accepted the Noble Animal, the humanist feeling about
such an animal, so frankly in jokes and casual talk' (ibid.). This is
why Boswell's *Life* – with its rich store of such material – provides
more than just an anecdotal supplement to the record of Johnson's
self-authored (and authorized) works. There is one passage in particular that Empson picks out as capturing precisely the way that
'dog' could evoke sentiments that Johnson suppressed – though
not without difficulty – when writing for the record. It occurs
during the account of his and Boswell's journey to the Western
Islands of Scotland and recollects their meeting with the Laird of
Col. They both tend to think of him, Empson remarks, as 'the best
example of the natural man' (sure in his instincts, fleet of foot,

physically the perfect specimen of his kind, living in harmony with nature). At the same time this feeling is yet another version of pastoral in so far as it brings in complex class-related attitudes. Thus the Laird was 'Johnson's social superior if not Boswell's and had been educated in England', despite which 'they insist on the romance of treating him as a savage' (*Complex Words*, p. 172). And it is here that the word 'dog' comes in, as a carrier for sentiments that perhaps neither of them (Johnson especially) could afford to spell out in more detail. 'You get a great deal of him in casual speech', Empson suggests, 'as a way to fix a line of sentiment', though the photographic metaphor seems to indicate not so much a quest for absolute clarity of focus but more a matter of achieving the right sort of distance to allow for some discreet blurring of the image. No doubt the word was felt as a joke, but one which 'brought in feelings that were important to them', feelings that required a certain vagueness of grasp if their other (more awkward or disturbing) implications were not to become too prominent.

The passage from Boswell is worth quoting lest it be thought that Empson is here piling an absurd weight of significance onto so simple and unassuming a word.

> Young Col told us that he could run down a greyhound; 'for' (said he) the dog runs himself out of breath, and then I catch up with him'. I accounted for his advantage over the dog, by remembering that Col had the gift of reason, and knew how to moderate his pace, which the dog had not sense enough to do (cited by Empson, p. 172).

'It is very winning', Empson comments, 'to see Boswell patting him on the head, as the better dog of the two' (ibid.). And Johnson's rejoinder takes the idea on a stage to the point where it seems precariously balanced between a good-willed (though superior) joke about the Laird and something much closer to the range of implications for which he took Fielding so sternly to task. As Boswell recollects: 'Dr. Johnson said, "He is the noblest animal. He is as complete an example of an islander as the mind can figure. He is a farmer, a sailor, a hunter, a fisher; he will run you down a dog; if any man has a *tail*, it is Col" ' (ibid.). What comes across here is about as full a version of the humanist and pro-dog sentiment ('man most triumphant of the animals') as the period has to show, or as Empson could find after much dogged hunting around in

literary and other sources. Clearly it was meant – and registered
with Boswell – as a piece of 'tremendous praise', such that 'while it
is ringing in our ears we can understand the generosity latent in
more satirical uses of the word' (p. 173). Yet those other uses would
have been much closer to Johnson's 'official' way of thinking, his
sense that the whole cluster of feelings around 'dog' was covert
propaganda for a thoroughly repugnant (irreligious or downright
blasphemous) creed. Thus it is only in Johnson's more off-the-cuff
remarks – when his doctrinal defences are down – that the word
could unexpectedly acquire this sense of wholehearted moral ap-
proval.

Of course there is a possible objection here: namely that we have
no evidence for all these subtle imputations of meaning, motive
and intent save another man's rendering of some offhand remarks,
written down many years later and from a viewpoint – Boswell's –
clearly more inclined to just that kind of vaguely disreputable
'pro-dog' sentiment. But this objection is not as damaging as it
seems since Boswell was himself sufficiently alert to both the range
of meanings carried by the word at that time – also borne out by
Empson's researches – and its resonance with other (no doubt more
complicated) feelings attached to it by Johnson. These latter are
most strikingly conveyed in his remark about Fielding as 'inventor
of the cant phrase, goodness of heart', a phrase that to his mind
signified 'little more than the virtue of a horse or dog'. Empson's
comment – that Johnson 'had to put *horse* beside *dog* to keep the
slang senses from rising at him' – is I think a good (if negative)
example of the intention-based semantics everywhere implicit in
his theory of complex words. It thus points the way to what Schiffer
frankly despairs of achieving, and what Davidson promises – but
fails to make good – with his talk of 'passing theories' as a minimalist substitute for knowing (or possessing) a language.[48] In the background of Johnson's performance with the word is a keen sense of
its various semantic possibilities, that is to say, the full range of
implications – cynical, satirical, mildly or wholeheartedly approv-
ing – some of which he needs to keep at bay if the pro-dog sentiment is not to push itself forward. Even so that sentiment remains
very active not only in his moments of unbuttoned usage – as when
praising the Laird for his doglike qualities – but also in those
passages where Johnson is on his dignity and anxious not to let
such meanings through. For there remains a certain normative
dimension to the word, an idea that Johnson must work hard to

resist since 'the most important point about the noble animal [Laird or dog] is that he is a reassuring object to contemplate'. Indeed, '[t]he fact that he can be patronized as no more than fundamental makes you think better of the race of man' (*Complex Words*, p. 173).

Swift is another case, Empson argues, where the dog-sentiment gives rise to some complicated feelings that would otherwise have lacked any adequate object to serve as a focus – or a means of release – for their charged and ambivalent nature. What makes his case different from Johnson's is the much greater intensity of these conflicts and the degree of cynical (or misanthropic) sentiment often involved. Also there is the constant sense that Swift's very sanity depends on excluding the 'rock-bottom' (sensualist) meaning of the word and maintaining a precarious balance between its rationalist and humanist implications. As a matter of explicit doctrine Swift inclined strongly to the rationalist side, the idea that human nature was inherently corrupt and redeemable only – if at all – through the exercise of reason and a constant effort of self-disciplined instinctual repression. (Hence – as commentators have remarked – his definition of man not as *animal rationalis* but, more guardedly, as *animal rationis capax*).[49] Still this imposed some considerable strain upon Swift's ability to keep his satires from falling into a Timon-like attitude of outright contempt for humanity and all its works. 'A surprising number of the great writers went mad', Empson reminds us, 'and most of them feared to; indeed, the more you respect reason the more you must fear the irrational' (*Complex Words*, p. 169). Thus 'dog' came in as a much-needed source of reassurance that there might, after all, be certain 'rock-bottom' instincts – shared with the animals – that did not necessarily equate with mere sensualism, sexual promiscuity, man's unregenerate ('animal') nature, and so forth. To this extent it could be viewed as somehow standing in for the human unconscious, for 'the source of those impulses that keep us sane', although at times 'they may mysteriously fail us as in drought'. So it was that the dog became an object of obscurely comforting sentiments even to one, such as Swift, whose thinking otherwise ran clean against the broader current of 'pro-dog' feeling. All the same, Empson thinks, 'Swift kept himself sane for as long as he did on secret doses of this feeling . . .; its goodhumour and humility are somehow at the back of and make endurable the most regal solvents of his irony' (p. 169).

However there comes a point – located by Empson in Book Four of *Gulliver's Travels* – where Swift's irony gets so completely out of touch with this complex of sentiments that it topples over into an attitude of total (near-insane) contempt for all merely human values and beliefs. Thus in the end Swift 'shie[s] away very decidedly from the humanist application', consigning Gulliver (his hitherto fairly reliable narrator) to a state of terminal disenchantment with humanity at large, and leaving only the Houyhnhnms – equine citizens of an inhuman rationalist utopia – as the last and surely desperate refuge for whatever 'positive' values are still on offer. As with Johnson, so here: the horse makes its entry as a creature required to carry the weight of anti-dog sentiment, or the idea that a line could be held between the acceptable (rationalist) and the unacceptable (secular-humanist) ranges of belief. But the strategy fails: 'Swift's ideal animal is the pointless "calculating horse"' (the horse as generous has a point all right, but not the horse as Cold Reason)' (p. 169). In other words both writers – Swift and Johnson – are forced back upon a complex series of verbal manoeuvres in order to prevent unwanted senses from forcefully 'rising up at them'. But where Johnson carries the process off with some measure of preserved (if hard-put) rational equanimity, Swift is driven to the point of undermining all possible grounds for mutual trust – or communicative uptake – between author and reader. In part this has to do with a conflict of feelings almost inescapable at the time, one that took rise – as Empson argues – from the Christian *versus* secular and, more specifically, the rationalist *versus* humanist contest waged over these and kindred words. Thus '[i]f one considers the richness of the intuitive poetry, of the emotional life the language took for granted, in the seventeenth century, it is surprising the eighteenth could make so much of its more narrow material, could base so much poetry on a doggish mock-heroic' (p. 169). But in Swift's case this narrowness also went along with a massively – indeed pathologically – heightened sense of the meanings that were liable to break through if the humanist sentiment was not held in check by a firm application of Christian-rationalist principle.

In fact this was a fairly late development, Empson suggests, since the early humanists – including liberal theologians like Erasmus – were able to convey such 'vaguely anti-Christian' (or at any rate suggestively heterodox) ideas without anything like that degree of knife-edge psychological conflict. Hence their consistent pedagogical stress on 'learning correct Latin' and 'reading classical literature in bulk': not exactly to put the message across 'under cover', but rather 'to put it across in a way that would allow the pupil to digest it within Christianity; as something that applied to life in a different way, rather than conflicted' (pp. 169–70). Here again Empson is invoking large dimensions of background feeling and belief but always with a view to understanding how they bear upon particular speakers (or authors) in particular contexts of utterance. And he does so, moreover, through a method that combines the resources of an intensionalist (logico-semantic) analysis with those of a sensitive and nuanced response to the *intentions* – at whatever unconscious or preconscious level – that can best explain the use of such complex words in just such particular contexts.

INTENTIONALITY, SEMANTICS, AND BELIEF-CONTENT

Of course it is still open for the sceptic to argue – with Schiffer – that this all presupposes what has yet to be proved, that is, the validity of an intention-based semantics that would make good its claim to reunite the various components that analysis had put asunder. But Empson already had an answer in the closing chapter of *Seven Types of Ambiguity* where he countered the prejudice of 'appreciative' critics, those who felt that poetry – or their own enjoyment of it – was somehow threatened by any such hard-pressed 'analytical' approach. To adopt this defensive posture, he thought, was to underrate both the resilience of poetry (the likelihood of its coming out unharmed, even enhanced, after the rigours of verbal exegesis) and the reader's capacity to gain likewise through a better, more conscious understanding of his or her intuiiive responses. No doubt it may be said – and with some justification – that 'the business of analysis is to progress from poetical to prosaic, from intuitive to intellectual, knowledge'; in which case the objectors can plausibly maintain some version of the stock ('intuitive' *versus* 'analytical') dichotomy. From this point of view '[a] poetical word is a thing conceived in itself and contains all its meanings; a prosaic word is flat and useful and might have been used differently'.[50] However – as Empson shows in *Seven Types* and with examples like 'dog' in *Complex Words* – there is a lot more going on in such 'flat', 'useful' and 'prosaic' sorts of language than is likely to be grasped by any reader who espouses the anti-analytical line. Thus '[t]hings temporarily or permanently inexplicable are not to be thought of as essentially different from things that can be explained in some terms that you happen to have at your disposal; nor can you have reason to think them likely to be different unless there is a great deal about the inexplicable things that you already know'.[51] (This will only seem a flatly contradictory or paradoxical statement if one takes it that the 'things' in question – poems or words – are somehow cut off from any larger background of shared linguistic understanding.) And by the same token – Empson maintains – there is less difference than might be supposed between the kinds of complexity that characterize poetic language and those that enter into our normal, everyday-communicative uses of language.

At any rate, '[w]hat is needed for literary satisfaction is not, "this is beautiful because of such and such a theory", but "this is all right; I am feeling correctly about this; I know the kind of way in which it is meant to be affecting me" '.[52] Which is also to say – as with the above-cited cases of Johnson and Swift – that we shall get much closer to an author's intentions (or to utterer's meaning in the everyday context) if we allow for the extent to which *all* communication involves both 'intuitive' and 'analytic' modes of understanding. For the alternative, as Empson sees it, is to introduce a false and harmful dichotomy, one that places poetry in a realm apart – accessible solely by grace of 'intuition', or maybe through privileged poetical tropes such as paradox or irony – while ignoring the sorts of complexity and richness that analysis can uncover in our everyday 'prosaic' uses of language. Besides there are other strong grounds, in literary criticism as elsewhere, for preferring an approach that has the courage of its analytic bearings and doesn't rely too much on unaided intuitions. After all, 'it often happens that, for historical reasons or what not, one can no longer appreciate a thing directly by poetical knowledge, and yet can rediscover it in a more controlled form by prosaic knowledge'.[53] Such is indeed the chief claim for Empson's method in *Complex Words*: that however 'prosaic' the terms of analysis (or the logico-semantic 'bits of machinery') employed, they none the less enable the critic to interpret meanings of the utmost subtlety and expressive power. Of course one may also need a good dictionary to hand – preferably one organized 'on historical principles', like the *OED* – in cases where the word has undergone some shift in its normal or acceptable range of usage. But Empson is not proposing anything like a Quinean experiment in 'radical translation', or a theory of scratch interpretation artificially divorced from all the usual contexts of natural-language communicative grasp. Rather he is attempting to show just how much 'analysis' may be required in order for speakers and interpreters (or poets and critics) to achieve the sort of 'intuitive' feel for language that enables them to communicate across and despite these shifts of historical-semantic perspective.

This applies just as much to everyday speech-situations of the type envisaged by Davidson, situations (that is) where the interpreter is confronted with some anomalous, deviant or otherwise puzzling expression that cannot be construed straightforwardly in accordance with his or her 'prior theory' as given by a knowledge of the language concerned.[54] For here also what must be in play – if interpretation is to succeed – is a sense of the various possibilities that exist (from metaphor, paradox and irony to malapropism and false analogy) whereby to explain why the utterance took that form and what (if any) meaningful construction to place upon it. This range of options is drastically narrowed with the switch to a Davidsonian dualism of 'prior' *versus* 'passing' theories. All that remains is the flat choice between a generalized language-competence incapable of handling any novel expression – any utterance that departs even slightly from the norm – and on the other hand a 'theory' whose field of application is indeed 'vanishingly small' since it is apt to work for just one speech-act in just one context of usage.[55] Nothing could be further from the kind of approach – the intention-based semantics in a full and non-reductive sense of that term – which Empson develops to a high point of methodological refinement. What is altogether lacking in Davidson's account is the idea of language as a complex field of interaction between the various possible senses of words, their structures of logico-semantic implication, and, above all, their capacity to articulate the subtlest nuances of speaker's intent through the differing degrees of prominence atttached to those senses in context.

This may all seem utterly remote from the problems that Schiffer encounters with his test-case sentence: 'Tanya believes that Gustav is a dog.' And yet – as I have argued – these problems might be seen as resulting not so much from genuine difficulties in the nature of linguistic comprehension but rather from the narrow, artificially restrictive approach that some philosophers standardly adopt in constructing such examples and debating their significance. In Schiffer's case this narrowness comes of a conviction that there exists no alternative to the kind of intention-based semantics that he once espoused but now believes to have failed on every relevant count. For the trouble with all such theories, he argues, is that they involve an atomistic (reductionist) treatment of meaning, truth, belief and speaker's intent which cannot re-establish any plausible link between the various component items. Still it may be argued that Schiffer is premature in reaching his end-of-the-road conclusion, that is, his view that 'analysis' is useless in whatever shape or form, and intention-based semantics a misguided endeavour on account of its analytic pedigree. For there does exist at least one alternative approach that manages to reconcile a powerful and sophisticated semantic theory with an approach that is fully alive to the range and multiplicity of utterer's meaning. This difference comes out most pointedly, as I have argued, if one compares Empson's chapter on 'The English Dog' with Schiffer's self-acknowledged failure to construe either Gustav's doghood or Tanya's belief as in any way susceptible of adequate analysis in semantic, belief-related, or intentionalist terms. 'If we are ever', Schiffer writes, 'to take seriously the idea that beliefs such as Tanya's are made complete by the presence within them of doggy stereotypes, we shall certainly require an articulation and treatment of that intuition that elevates it into a hypothesis worth considering.'[56] And since, on his account, no such hypothesis can possibly be sustained – given the inscrutability of reference and the absence of sufficient (analytical) criteria for individuating either dogs or beliefs about them – therefore the entire project collapses into manifest incoherence. For there is just no way (so the argument runs) that belief-contents can be hooked up, *via* a theory of propositional attitudes, to an adequate grasp of speaker's intentions with respect to such items of belief.

In his later work Empson conducted a full-scale campaign against the doctrine of the 'Intentional Fallacy'. This was the idea – erected into a high point of principle by orthodox New Critics like W.K. Wimsatt – that we cannot (or should not) fall back on vague talk of authorial 'intention' as a substitute for the more exacting demands of textual close-reading or scrupulous attentiveness to the sacrosanct 'words on the page'.[57] This argument had three main prongs: first, that such intentions could only be matter for conjecture, belonging as they did to a private realm inaccessible to any interpreter; second, that no amount of 'evidence' (for example, from an author's working notes, journal entries, or recorded statements of intent) could ever be decisive since the poem might always quite possibly have turned out to mean something other than what they intended; and third, that such appeals were in any case irrelevant since a good poem would communicate its meaning without the need for such extraneous sources of information.[58] So there could be no warrant for intentionalist talk, amounting as it did either to an admission of defeat on the critic's part or else to a kind of special pleading for poems that had failed to get their meaning across in a sufficiently focused or concentrated verbal form. Hence the favoured New Critical rhetoric of 'irony', 'paradox', 'tension' and so forth, joined to the idea of the poem as a 'verbal icon' whose structure was objectively *there* on the page and in no way dependent on a merely subjectivist appeal to authorial intent.[59]

When arguing directly against this doctrine – as so often in his later essays and reviews – Empson tends to adopt a plain commonsensical (if sometimes exasperated) tone. Thus it just *is* the case, self-evident to all but these 'bother-headed' theorists, that human beings are constantly involved in the business of figuring out other people's intentions; that we could not make a start in learning to talk and communicate unless we were able to bring this off with a fair degree of success; and that any argument to the contrary – such as Wimsatt's – can only show the critic to be in the grip of some dogmatic (most likely theological) belief about the limits of human reason or intelligence. The following passage – from a 1955 review of Wimsatt's book *The Verbal Icon* – is a fair sample of Empson's way with what he saw as a deeply misguided, obscurantist, and at times morally pernicious doctrine.

> Among the first things a baby has to learn, and if it can't it's mad, is that other people exist; if it couldn't feel 'Mum's cross' and so forth before it learned to speak, then it couldn't learn to speak. Estimating other people's intentions is one of the things we do all the time without knowing how we are doing it, just as we don't play catch by the Theory of Dynamics. Consider the Law, which might be expected to reject a popular fallacy; it recognises amply that one can tell a man's intention, and ought to judge him by it. Only in the criticism of imaginative literature, a thing delicately concerned with human intimacy, are we told that we must give up all idea of knowing his intention.[60]

This is of course to state the case in a highly generalized (even, some may feel, a rather crude and knock-down) polemical form.

After all, there are genuine – not merely 'bother-headed' – problems about the doctrine of *mens rea* in law, that is, the extent to which parties can be held fully responsible for words, actions and consequences that may admit of no clear or unambiguous construal in terms of speaker's or agent's intent. All the same Empson does have a point when he argues that we could not even raise such issues – whether in the legal, the literary, or the everyday social-communicative context – if we did not possess at least a fair working grasp of how intentions are normally or typically manifest in human speech and behaviour.

In fact the passage goes on to make this point in a different (though related) context. Thus:

> Mr. Wimsatt includes an article on 'Poetry and Christian Thinking', written seriously as a Christian, and feels he has to admit that Biblical criticism has been 'implicitly intentionalist' because it has 'dealt with inspiration'. But surely the obvious, as well as the ancient, opinion is that there can indeed be occasions when the author is *hardly* answerable for his intention – when he is inspired. It shows great staying-power, I think, to hold the puzzle so firmly the wrong way up.[61]

Clearly Empson is not talking about divine inspiration – the word of God vouchsafed to some spiritual elect – since there is plentiful evidence elsewhere in his work that he regarded such beliefs as inherently deluded and a major cause of man's inhumanity to man.[62] Rather, he is suggesting that poets – and for that matter speakers – may sometimes say more than they mean to say, or come up with expressions more complex and obscurely evocative than can well be thought to have formed a part of their conscious or deliberate intent. Interpreters down the ages have found various ways of handling this sort of case, whether in religious (inspirational) terms, through Romantic talk of 'genius' – its quasi-secularized equivalent – or again, by invoking the Freudian 'unconscious' as a handy refuge when they propose some reading that strains the limits of a straightforward intentionalist account.

Empson is far from rejecting this latter line of argument since it is one that he adopts at various points in *Seven Types*. He does so mainly with a view to heading off the predictable objection that (for instance) a Christian-devotional poet like George Herbert cannot possibly have *meant* his poem to be read as a series of self-tormenting reflections on the hideous doctrine of God's 'satisfaction' through the suffering and death of Christ.[63] Thus the concept of 'intention' – or its range of application – may well need stretching in order to accommodate such cases. But it is also Empson's firm belief that one cannot just cast it aside, like the New Critics, and treat the poem as a 'verbal icon', an inwrought structure of paradoxical meanings entirely cut off from the poet's intentions, no matter how complex or contradictory. For the result – as he sees very clearly in Wimsatt – is to place what amounts to an orthodox ban on any reading that would raise serious questions first about the content of Christian doctrine, and second about that doctrine's effect upon Herbert as he struggles to accept and redeem its 'appalling' implications. To treat these latter as 'paradoxes' – rather than a series of massive and mind-wrenching contradictions – is to take them on faith as giving access to a wisdom (an order of revealed religious truth) beyond the grasp of mere unaided human intellect. It thus underrates both the sheer expressive power of Herbert's poetry and the extent to which even the most extreme cases of seventh-type Ambiguity can only make sense – only register as such – against a normative (maybe 'prosaic') background of intelligible motives and beliefs. Moreover, what makes 'The Sacrifice' so remarkable a poem is our sense of the extraordinary strain that these contradictions place upon the language – the semantics and the logical grammar – of *each and every stanza* in a long narrative sequence.

Thus Jesus figures at the end 'as the complete Christ; scapegoat and tragic hero; loved because hated; hated because godlike; freeing from torture because tortured; torturing his torturers because all-merciful; source of all strength to men because by accepting he exaggerates their weakness; and, because outcast, creating the possibility of society'.[64] But again we shall have little sense of these powerful and haunting contradictions if we opt instead – like the New Critics – to view them as 'paradoxes' somehow intrinsic to the nature of Christian faith and hence as simply off-bounds for the purpose of rational prose explication. It is not hard to see why the adepts of that movement took exception to certain aspects of Empson's work – his indulging in the various 'fallacies' (or heresies) of paraphrase, intentionalism, and the like – while professing admiration for his brilliance and acumen as a close-reader of texts.[65] Nor is it hard to understand Empson's often somewhat rattled and cantankerous response at being taken on board as a precocious maverick, one whose best insights had to be sifted from his regrettable lapses into just those sorts of naive error. For he saw clearly enough what these critics were at: picking up whatever suited their purpose in the way of 'deep' (paradoxical) import while cutting out whatever might count against it from the standpoint of a method – an intention-based semantics – not thus committed to respecting the requirements of an orthodox religious creed. In 1951, writing for a *Kenyon Review* symposium ('The Verbal Analysis') alongside several New Critics, Empson felt able to note this tendency without much sense of alarm. Thus: 'I should think indeed that a profound enough criticism could extract an entire cultural history from a simple lyric, rather like Lancelot Andrewes and his fellow preachers, "dividing the Word of God", who were in the habit of extracting all Protestant theology from a single text.'[66]

Nothing much to worry about here, it seems, even if 'it is not really a convenient way to teach cultural history'. But over the next two decades Empson grew more convinced of the moral and intellectual damage being done by this neo-orthodox movement in literary criticism. Thus his main efforts after *Complex Words* were devoted on the one hand to a polemical defence of secular-humanist values, and on the other to 'rescuing' various authors – Shakespeare, Donne, Milton, Fielding, Coleridge and Joyce among them – from the treatment they had received at the hands of these pious exegetes.[67] My chief concern here is not so much with the detailed literary-critical aspects of Empson's work as with the background theory of interpretation – the intention-based semantics – worked out in the earlier book. All the same (and the point is worth repeating) there is a strongly marked normative or ethical dimension to his theory of complex words, a constant appeal to those uses of language – or modes of communicative utterance – that provide a kind of pattern for the process of human understanding in general. So the book is better seen as having laid the groundwork, the logico-semantic 'machinery', for just that kind of explicitly secular (rational-humanist) approach that Empson adopts in his later, more polemical essays. Indeed there is a whole implicit anthropology – a theory of human needs, desires, values, and mutual obligations – bound up with his deeply naturalistic account of language and interpretation. It emerges most clearly in the chapters on keywords like 'sense', 'honest', and 'dog', words that on occasion may be twisted into bearing a range of malign, cynical, or anti-humanist sentiments, but which also – more typically – carry along with them a certain 'humour of mutuality', an appeal to those elements of shared understanding that alone make communication possible.

Such is the root feeling about 'dog', Empson thinks: 'that there is a sweetness or richness in the simple thing, that to cut yourself off from it would be folly, that it holds *in posse* all later values; also the fact that you need to remind yourself of it may show how encouragingly far it has been left behind' (*Complex Words*, p. 170). It is in remarks like this – scattered throughout *Complex Words* – that Empson declares himself strongly in favour of a naturalized ethic with substantive implications for linguistics, interpretation-theory, and the human sciences in general.[68] Thus the passage continues: what is essential here (that is, about 'dog' applied to human beings as a term of praise) 'is that you can start building yourself into a man, and not hate yourself, on the basis of being that kind of animal; the trouble about Evolution was that one could not feel the same about monkeys' (p. 170). Empson has some interesting background detail on the topic of monkeys (also apes and orang-outangs) and the sorts of anxiety their behaviour provoked among eighteenth-century intellectuals, Johnson and Boswell included. (Boswell found the resemblance between humans and monkeys 'sadly humbling'; Lord Monboddo, who had brought an ape back from Africa, 'only dared to say what others had suspected'; and Dr. Johnson for his part had to agree that Monboddo 'was not a fool'; ibid.). At any rate there is plenty of evidence that the evolutionary hypothesis – or the idea of man as in some way related to the rest of the animal creation – was already well advanced in the century or so before Darwin laid it out in scientifcally elaborated form.[69] The main question was just *which* animals might lie closest to man on the scale of intellectual and moral evolution to date. And it is here – Empson argues – that the dog came in as an alternative (humanly more acceptable) substitute for the idea of mankind as next in line from the monkeys and apes.

Again Swift provides something of a touchstone for this complex of attitudes on its negative (cynical or anti-humanist) side. In Book IV of *Gulliver* he divides the world between the Houyhnhnms – the purely rational 'calculating horses' – and the yahoos, filthy monkey-like creatures supposed to embody all the worst, most depraved attributes of fallen humanity. But this was an evasion of the issue, Empson thinks, since it left no room for that other, more hopeful or generous complex of feelings that had grown up around the name and nature of 'dog'. (Of course he is not so much talking science here – at least not 'science' from the standpoint of our present-day best, most adequate theories – as reconstructing what seems to have emerged at that time as a mixture of advanced proto-evolutionary ideas and other, to some extent 'folk-psychological' beliefs.) Thus 'Swift might fall back on the Houyhnhnm in accepting this about the yahoo, but that was a refusal of humanity; the only real animal to use was the dog' (*Complex Words*, p. 170). For the point about dogs, considered in this light, was that although less 'intelligent' than the apes and monkeys – less evolved toward some sorts of rational-seeming behaviour – they could still be felt to manifest other, more humanly appealing traits (fidelity, mutual reliance, instinctive generosity, the basic social virtues) that gave grounds for hope that the process was headed in the right direction. 'And indeed', Empson writes,

> it is comforting to reflect that this apparent evasion was to a great extent the truth. Whatever the spiritual quality we dislike in monkeys may be, there is no positive evidence that our common ancestors shared it; and even if they did, for a much greater period of time they were straightforward mammals like a dog or a squirrel. Even our ancestral reptile, in the pictures, is made to stand up on its legs and look about it like a puppy. There is a curious agreement, at any rate, that if we are animals this is the kind of animal we would like to be (*Complex Words*, p. 170).

Of course there is a sense – a scientific sense – in which we must now view this 'evasion' as really just that, an avoidance of the truth about mankind's proximity to the apes and monkeys on the scale of evolutionary descent. So when Empson talks of animal 'kinds' and the affinities between them – including the human sympathy with dogs as representing 'the kind of animal we would like to be' – he cannot be taken as advancing this claim from a strictly scientific (post-Darwinian) perspective. On the other hand he can appeal to natural kinds – to apes, monkeys, dogs, and squirrels – as having offered a range of candidate ancestors whose traits, both physical and temperamental, rendered them more or less humanly appealing.

One can see here one reason why philosophers like Schiffer are driven into postures of extreme and self-disabling scepticism by fixing the requirements for natural-kind reference ('Tanya believes that Gustav is a dog') at too high a level of scientific expertise.[70] Thus Schiffer makes his case for the indeterminacy of reference, meaning, and belief-content on the grounds that Tanya will most likely not be in possession of the relevant specialized knowledge (evolutionary, genetic, species-taxonomic and so forth) to know for sure that Gustav is a *bona fide* specimen of the kind 'dog', or to give the interpreter sufficient assurance that she is using the word to pick out *all and only* members of the canine kind. From which it follows, on Schiffer's account, that her belief about Gustav lacks determinate content and likewise any intention-based semantics (or analysis in terms of propositional attitudes) that naively assumes her to know full well what is meant by the sentence 'Gustav is a dog'. Of course Empson's purpose in *Complex Words* is not to offer a philosophically elaborated theory of reference, meaning and intention that would meet such sceptical arguments point for point on technical grounds of their own choosing. Indeed he is frankly impatient with much of what he sees as the 'bother-headed' tendency – among philosophers and literary theorists alike – to raise pseudo-problems which get in the way of any adequate (humanly responsible) approach to language and interpretation. All the same *Complex Words* is, among other things, a work of philosophical semantics and one that points a way through and beyond the kinds of hyperinduced sceptical doubt that Schiffer so exhaustively articulates. Moreover it offers a range of arguments – about meaning, intention, belief, natural kinds, the structure and content of propositional attitudes – which do have a bearing on these issues in present-day philosophy of mind and language, whatever Empson's reluctance to discuss such matters in the abstract.

Thus when Johnson and Boswell praise the Laird of Col for his supremely natural or instinctive (dog-like) qualities they are picking up a line of sentiment – a complex of animal-related feelings – that somehow communicates despite and across all the intervening shifts of semantic, cultural, and (not least) scientifically-accredited knowledge. That is to say, there are other criteria for natural-kind membership than those having to do with the idea of doghood as fixed – or beliefs about dogs as individuated – by an expert knowledge of their genetic constitution, phenotypal characteristics, or suchlike. In Tanya's case there is the acquired capacity (both cognitive and linguistic) to pick out candidate exemplars of the kind across a fairly wide range of otherwise diverse breeds, shapes, sizes, colours, or behavioural traits. And in doubtful or borderline cases she can always defer to what Putnam calls the linguistic 'division of labour', the assurance that there exist other people –

experts in the relevant domain – who could settle the issue if called upon.[71] Of course there was no such recourse for those eighteenth-century speculative types who wondered (sometimes fretfully, sometimes with a sense of new-found optimism) about the possible relation between man and ape or man and dog. But despite this lack of a more developed scientific understanding they were clearly working on a set of assumptions that enabled them – and still enable us – to render such feelings intelligible. Thus:

> Monboddo's theoretical writing, though certainly inspired by the problems of the ape, is very insistent that the interesting cases are some entirely human creatures who merely happen not to have discovered speech; these people (about whom he has elaborate and quite false information) set an interesting problem to the linguist and philosopher, but it is very unfair to them when people mistake them for apes (*Complex Words*, p. 172).

It is worth noting that there is an unfinished work of fiction in the Empson *Nachlass* – entitled *The Royal Beasts* and posthumously edited by John Haffenden – which raises similar issues about the ethical rights and prerogatives of a borderline species (apparently talking orang-utans) who prefer to be classified as non-human since that way they will enjoy protected status, not have to pay taxes, and generally be better treated.[72] Moreover we are clearly meant to admire these creatures for their tolerance, wisdom, and large-minded grasp of the ethical issues involved, as well as for the shrewdly calculating intelligence that leads them to adopt this policy. Thus the story works out as a kind of inverted variation on Swiftian themes. It offers an evolutionary fable where the satire is more benevolent – more generous and hopeful – to the extent that it allows for elements in animal nature that bring out the potential best (rather than the taken-for-granted worst) in the counterpart world of human motives and interests.

The most important point here, as in *Complex Words*, is the scope that it offers for an interpretation-theory which need not (like Quine's) start out from a basis of presupposed total ignorance about the sorts of understanding – or knowledge-constitutive interest – that make communication possible.[73] After all one could argue that Empson's fictive thought-experiment in *The Royal Beasts* is no more extravagant or wildly counterfactual than Quine's idea of 'radical translation' as somehow taking place between the field-anthropologist and the native informant, each of them working absolutely from scratch, with no knowledge whatsoever of the other's language, customs, beliefs, intentions, needs or desires.[74] Any theorist who follows through on this programme is sure to end up, like Schiffer, in a position of doubting – contrary to all the evidence – that we could ever have adequate grounds for imputing determinate meanings or beliefs. Of course it may be countered that Empson's theory is premised on a set of normative (indeed anthropological) assumptions – ideas about human nature, about man's relation to other animals, about the role of evolution in fostering values of mutual reliance and socialized communicative exchange – which simply fail to meet the sceptic's question on each of the main points at issue. Thus it has long been supposed (by philosophers from Kant to Husserl and also by many in the recent 'analytic' mainstream) that any adequate theory of mind, language or interpretation will need to slough off its residual attachment to anthropological (human, all-too-human) values and beliefs. However, there are others – among them proponents of a naturalized epistemology – who have come to see this as a hopeless endeavour for just the sorts of reason that Schiffer adduces in his failed attempt to make the programme stick.[75] That is, it will always at some point run aground on the problems that bedevilled logical empiricism and that have cropped up again with each new attempt to establish a formalized (truth-theoretic) semantics for natural language.

Such, as we have seen, is Davidson's nagging problem: the failure of a purely formal and recursive analysis of truth-conditions to offer any guidance when it comes to more substantive issues of truth, meaning, and interpretation.[76] However, the prospect is no better for those currently fashionable 'post-analytic' approaches that take a lead from Quine's 'Two Dogmas' and opt for a radical empiricist outlook relieved of such otiose 'metaphysical' commitments. For to naturalize epistemology in the Quinean way is to buy out of the logical-empiricist dilemma (along with all its vexing dualisms: form/content, analytic/synthetic, concept/intuition, etc.) but only at the cost of relativizing truth to some notional totality of beliefs-held-true by this or that community at this or that time.[77] Thus it leaves no room for anything like an intention-based compositional semantics of the type once sought by Schiffer and most convincingly achieved – as I have argued here – by Empson in *Complex Words*. For the three chief requisites of any such account are that it should (1) provide some means of individuating the objects or contents of belief; (2) relate these to utterer's meaning *via* a semantic (intensional) analysis linked to a theory of propositional attitudes; and (3) explain how speakers can communicate particular (perhaps non-standard) meanings and beliefs in particular (more or less well-defined) contexts of utterance. None of these requirements can possibly be satisfied if one adopts Quine's holistic approach to issues of truth, meaning, and belief. Item (1) clearly has to go since – on the thesis of ontological relativity – there is just no way of individuating objects or belief-contents except in relation to the entire 'web' or 'fabric' of other such commitments at any given time. Item (2) drops out for much the same reason, that is, Quine's attitude of wholesale scepticism with regard to the claims of analysis (or compositional semantics) as a method for assigning content to beliefs through an account of their relevant truth-conditions. (These, after all, are among the residual 'dogmas of empiricism' which his essay sets out to demolish.) And item (3) emerges in yet worse shape since on Quine's account any novel utterance – like any discrepant or anomalous finding in the natural sciences – can always be finessed by some suitable reinterpretation, some pragmatic 'adjustment' that brings it into line with the existing totality of beliefs.

'RADICAL TRANSLATION': MANAGING WITHOUT

However there exist some well-developed alternatives to both these Quinean 'radical' positions, that is, the strain of radical empiricism that ends up by espousing a doctrine of wholesale ontological relativity and – following from this – the idea of (so-called) 'radical translation' as a scratch theory adopted for lack of more adequate interpretive resources. On the first point – as Kornblith and others have argued – it is wrong to suppose that a 'naturalized' epistemology must always issue in a sceptical, a relativist or conventionalist attitude *vis-à-vis* those objects or contents of belief that figure in our various ontological commitments.[78] Indeed there is strong support – both from causal-explanatory theories in philosophy of science and from the causal theory of reference developed by philosophers like Kripke and Putnam – for the view that what enables us to pick out such objects is their belonging to types, categories, species or natural kinds whose salient (individuating) properties or features are just those that serve to justify such (otherwise arbitrary) practices of classification.[79] So it is, Kornblith argues, 'that our inductive inferences are tailored to the causal structure of the world, and thus that inductive understanding of the world is possible'.[80] And this in turn provides a basis for challenging the second of Quine's 'radical' conclusions, his idea that any adequate (philosophically cogent) account of linguistic understanding must necessarily start out – like the 'radical translator' – from a position of total ignorance concerning the interlocutor's language, beliefs, customs, worldview, ontological scheme, and so forth. For if epistemology can be naturalized in the way that Kornblith more plausibly suggests then there is simply no reason – obdurate scepticism aside – to regard this as anything other than a *reductio ad absurdum* of the sceptical-relativist case.

It also indicates what is wrong with the other (Davidsonian) alternative candidate-theory, that which invokes the Principle of Charity as a means of maximizing the truth-content of any sample utterance and thus, supposedly, enabling both parties to come out 'right in most matters'.[81] This argument fails for the reason that Empson makes clear in his discussion of Monboddo and the orangutan that there may be instances where beliefs are formed on the basis of 'elaborate but quite false information', and yet where those beliefs are put across with some success and indeed – in this particular case – 'set an interesting problem to the linguist and philosopher'. Monboddo's mistaken idea was that these were entirely human creatures who for some reason had not acquired the faculty of speech. Thus he feels it to be 'very unfair to them [the orangutans] when people mistake them for apes' (*Complex Words*, p. 172). But his error is understandable – Empson thinks – if one takes account of the rich and complicated background of feelings at work. Nor is it in any way coincidental that this point about the sometimes complex relation between reference, meaning and speaker-intent comes up in a context – that of animal species and natural kinds – that associates so readily with the case for a naturalized epistemology. For what Empson considers so crucial about the complex of proto-evolutionary meanings around words such as 'dog' and 'ape' is their capacity to convey a whole range of sentiments – even 'compacted doctrines' – that would otherwise have found no adequate expression in the official (theologically sanctioned) discourse of the time.

Here again we are much closer to Papineau than to Davidson, that is to say, more within the range of a 'principle of humanity' that allows for *false yet humanly intelligible* beliefs than a wholesale face-saving 'principle of charity' that maximizes truth as a general rule and thus leaves no room for such crucial epistemic distinctions. Hence Papineau's moral: that 'if we can *understand* why some alien people should accept the judgements our interpretation attributes to them, those attributions should count positively in favour of our interpretation, even if the judgements in question are false'.[82] Thus Monboddo was definitely wrong on certain points, among them this point of species-attribution and his mixing up the evolutionary issue with those other (to him more interesting) cases of 'entirely human creatures who merely happen not to have discovered speech'. Still he managed to raise some important questions not only with regard to the then-nascent disciplines of ethnology, anthropology, and natural history but also – as Empson remarks – for linguists and philosophers with a fairly open mind on this issue of our relation to the other animals.

Let me cite another passage from *Complex Words* which brings out this deep connection between the normative aspects of language-usage, their strongly marked ethical dimension, and the view of human nature (the implicit anthropology) assumed to underlie such manifestations of the social-communicative lifeworld. 'I doubt', Empson writes, 'whether there is a case for saying that the actual sight of apes was what startled Europe into a change of thought.' After all,

> [t]he time would have to be earlier than Descartes, whose assertion that animals are machines but men are not was important mainly because it gave great publicity to the question, together with an evidently wrong answer. Christianity requires a sharp distinction between the creatures with eternity before them and those without, and given the mechanistic scientific idea a rigid mechanist view of animals was the only one that fitted comfortably. This would be enough to raise a new puzzle about the relations of men and animals, seeing that it is difficult to take a crudely mechanistic view of a dog you go hunting with; so that if we are to consider the apes a crucial factor we must show that they put ideas into the mind of Descartes; and there were reasons to make him think that way even if he had never heard of them (*Complex Words*, p. 172).

This seems to me a good example of the background 'machinery' at work in Empsonian intention-based semantics. That is, it bears witness to his achievement in successfully joining together those components of a naturalized, speaker-relative, logico-semantic and context-sensitive theory that philosophers have lately either failed to integrate or, like Schiffer, seemed intent upon putting asunder. Also it shows (*contra* Davidson) the importance of limiting charity to a generalized assumption that speakers will most often *make sense* by their own best-available scientific, cultural, or linguistic lights, rather than counting them presumptively *right* – in possession of the relevant truths – on an across-the-board charitable basis. For unless one builds in a defeasibility-clause – some allowance for their having quite possibly gone wrong through ignorance, partial information, or predisposed habits of belief – then the principle becomes either altogether vacuous or apt to produce any number of erroneous results.

5

Realism, Truth and Counterfactual Possibility: Thought-experiments in Science and Philosophy

ANTI-REALISM: THE WORLD WELL LOST?

Much has been written on the topic of thought-experiments in the natural sciences, from Galileo's classic *a priori* proof that falling bodies are subject to a uniform rate of acceleration irrespective of their mass, to Einstein's debates with Niels Bohr about the completeness of quantum-mechanical theory, and – more recently – J.S. Bell's counter-intuitive results with regard to particle spin and quantum action-at-a-distance.[1] What they all have in common is a readiness to accept that conclusions arrived at through the exercise of thought in its critical-speculative mode may actually require certain modifications to our concept of physical reality, sometimes to the point of drastically revising what counts as a viable candidate theory.

For some sceptically inclined philosophers of science – Quine among them – the lesson to be drawn is that of ontological relativity, i.e., that observations are always to some extent 'theory-laden' and theories always 'underdetermined' by the best evidence to hand.[2] In which case, according to Quine, we had better accept that there is no privileged ontological scheme that would allow us to distinguish (say) centaurs, Homer's gods, mathematical sets or classes, and brick houses on Elm Street in point of their 'reality' or existential status. For others it suffices to establish the case for an outlook of wholesale anti-realism that would simply have done with such vain attempts to fix the relevant boundary-markers. However, these solutions are purchased at considerable cost since they introduce a whole raft of associated problems – referential opacity, incommensurable paradigms, radical meaning-variance,

154

and so forth – which have proved yet more intractable.[3] Unless, that is, we follow Richard Rorty in declaring them simply *dépassé* for anyone willing to switch language-games or regard 'truth' as just an honorific label attached to whatever is currently and contingently 'good in the way of belief'.[4] But there seems, on balance, rather little to commend any theory that makes a chief merit of its own total failure to explain what distinguishes (say) astronomy from astrology, medical science from witchcraft, or oxygen from phlogiston as elements involved in the chemical process of combustion.

As so often with Rorty we can learn a good deal by going back to those problems he wants us to forget and then looking more closely at other (non-dismissive) efforts to address them. What they all have in common – on Rorty's account – is a misplaced anxiety about realism, that is to say, about the existence of an external (mind-independent) physical world whose nature, properties, causal dispositions, microstructural attributes, and so on, it is the business of science to describe and explain. Since this is clearly impossible – as any 'facts' thus described will always be subject to the inbuilt circularity that comes of their belonging to some language, discourse, or 'conceptual scheme' – we might as well simply give up worrying and abandon all such hopeless endeavours to redeem the notion of truth-as-correspondence.[5] I shall argue, on the contrary, that this is a false dilemma and one whose continuing hold upon philosophers in the analytic (and 'post-analytic') tradition can be traced back to the unresolved – and *on their own terms* unresolvable – dichotomies of old-style logical empiricism. Rorty himself remains stuck at that point, ceaselessly rehearsing the standard objections, such as the above circularity argument, while rejecting any stronger version of the realist case that would offer an alternative to his own pragmatist escape-route. And the same can be said of those other, more fretful types who remain in varying degrees dissatisfied with the Rortian solution but who can see no way beyond the sorts of dilemma that arise within the logical-empiricist paradigm.[6]

The alternatives in question are fairly well known so there is no need here for any lengthy review. Sufficient to say that they all strike Rorty as symptomatic cases of a backsliding tendency toward realism whose advocates appear to have learned nothing from that whole sad history of failed endeavour. Thus for instance he gives short shrift to the causal-realist theory of naming and necessity propounded by Kripke, Donnellan, and the early Putnam.[7] On this account certain words – paradigmatically proper names and natural-kind terms – have their reference fixed by an act of 'rigid designation' whereby they must thereafter be taken as applying to *just that* referent despite any later shifts in the range of criteria, descriptive predicates, or identifying properties which (on the alternative descriptivist theory) alone make it possible to pick out the object in question. The great advantage of this causal-realist approach – so Kripke claims – is that it avoids all those problems of ontological relativity and radical meaning-variance that result from applying the descriptivist model in contexts where the criteria for valid usage of certain terms (such as 'mass' in the history of physics from Aristotle to Galileo, Newton, and Einstein) have been subject to drastic revision. For it can now be seen that such terms have an underlying continuity of reference despite and across the various changes in what is taken to constitute an adequate description or definition.[8]

Thus the substance 'gold' may be identified as that to which the rigid designator 'gold' has always necessarily applied, even though its definition has shifted over time from 'a yellow, ductile metal that dissolves in *aqua regia*' to 'the one and only metallic element with atomic number 79'. So likewise with 'water' – defined now with reference to its molecular constitution H_2O – and other such terms ('atom', 'gravity', 'species', 'electricity', and indeed 'natural kind' among them) which have doubtless undergone some radical changes of descriptive or theoretical usage, but which can still be understood as picking out the same objects of continuing enquiry. More than that: they *must* be so understood if the history of science is to make any sense in terms of our knowledge of the growth of knowledge or our capacity to grasp what has actually changed in the sorts of understanding that science provides.[9] For on the descriptivist account – at least when pushed to its sceptical-relativist conclusion as by philosophers like Quine and Kuhn – there is simply no explaining how (for instance) Faraday achieved a real advance over previous investigations in the field through his experiments to show that a range of apparently disparate phenomena – electro-magnetic, voltaic, and static discharges – were in fact different manifestations of the same elemental phenomenon, namely that of electricity.[10] And this applies also to later developments, for example, the definition of electrical conductivity in terms of the passage of free electrons through certain sorts of material. For it remains the case that any progress toward a more adequate (depth-explanatory) account of electricity and suchlike phenomena must depend upon the continuity of reference that enables scientists – and philosophers or historians of science – to pick out relevant samples of just that kind.

As I have said, Rorty has no time for such arguments, or at any rate shows less patience with them than with most of those entertaining errors and illusions that make up the richly varied story of philosophy to date. For him they represent just another, particularly retrograde symptom of the age-old 'metaphysical' desire to discover some privileged conceptual scheme – some correspondence-theory or ultimate 'God's-eye view' – which would enable us to hook up our beliefs and theories with whatever it is in 'reality' that renders them objectively true or false. I think that this comes of Rorty's fixed assumption that the new causal realists are just old logical empiricists under a different name, and hence that they cannot have anything of interest to say that won't give rise to the same intractable problems. All the same his verdict has at least some justification in so far as theorists like Kripke and (the early) Putnam tend very often – though not always – to construe the causal link as a matter of *de dicto* rather than *de re* necessity. Thus they take it that terms acquire reference through an inaugural act of ostensive or stipulative naming, and that this reference is then passed on through the 'chain' of transmission by a process whose causal nature is defined as a matter of strictly linguistic necessity.[11]

To be sure, there are some passages in Kripke and Putnam – most often in the speculative or thought-experimental mode – where the notion of a causal-explanatory theory is extended to encompass scientific issues outside this rather narrowly semantic domain. Such issues are plainly in view when Putnam conducts his famous 'twin-earth' experiment with the two kinds of substance (H_2O and XYZ) both identified as 'water' by denizens of the respective – actual and imaginary – worlds, but only one of them answering to our use of 'water' as defined by its this-world molecular composition.[12] And there are other passages in their work that would seem to have substantive – non-trivial or extralinguistic – implications for an adequate causal-explanatory theory of naming, necessity, and natural kinds. But on the whole discussions in this area have tended to operate at a large and, I think, a methodologically disabling remove from the kind of detailed work in philosophy of science carried on by causal realists such as Wesley Salmon and

Roy Bhaskar.[13] Perhaps this is one reason why Putnam embarked upon his 20-year trek from a fairly robust (if underdeveloped) realist semantics to a theory of so-called 'internal realism' – realism internal to this or that frame of reference – which claims a Kantian pedigree but often sounds more like a minor variation on Rorty's neopragmatist theme.[14] Thus, in his case at least, there has turned out to be some warrant for Rorty's long-held view that eventually these realist types (or the less dogmatic among them) would surely see the light and come back around to his own way of thinking.

However we should not be too quick to accept that this is the end of the story. For there are, as I have suggested, alternative resources – in philosophy of science, philosophical semantics, and other related disciplines – that point a way beyond this supposedly inescapable pragmatist outcome. In particular I am thinking of two recent books by James Robert Brown which bring together some striking evidence and arguments in support of the realist case.[15] Brown's main interest is in thought-experiments and their capacity – on the face of it something quite remarkable – to mirror and sometimes to pre-empt those findings that can otherwise be arrived at only by dint of empirical observation or research. The following passage is very much to the point as regards both the strengths and the weaknesses of a causal theory of naming and necessity in the Kripke/early Putnam mode. Thus Brown:

> The causes in the causal theory of knowledge have always been thought of as efficient causes of a completely physical sort. (Photons come from the tea cup, interact with the rods and cones in the eye etc.) Let us expand the causal theory of knowledge to include the causal powers of abstract objects. Thus, we will suppose there are abstract objects – numbers, values, properties, laws of nature – and that these things causally interact with us, though not in any sort of physical way. Following Gödel, we can say that (in some special cases) we 'perceive' these independently existing abstract objects just as surely as we perceive tea cups. This sort of perception – often called intuition – has much in common with ordinary experience; but since it does not involve the ordinary physical senses, any knowledge deriving from it is justly called a priori. Let us so baptize them; if knowledge is based on sensory experience involving a physical causal connection, then it is a posteriori; if it involves an abstract causal connection by-passing the physical senses, then it is a priori.[16]

Clearly this distinction falls out very differently from that which enables Kripke and Putnam to stake their claim for an order of *a posteriori* necessary truths. For these latter have the curious property that they are thought of on the one hand as products of empirical discovery (hence as in no sense *a priori*), but on the other hand as truths whose validity is somehow guaranteed *in advance* by their pertaining to proper names or natural-kind terms whose reference is fixed – and causally transmitted – in such a way as to predetermine their correct usage.

This certainly gets over some of the problems with descriptivist theories which – as we have seen – open the way to wholesale ontological relativism or radical meaning-variance across diverse 'incommensurable' paradigms. But it does so at the price of fixing reference in a purely stipulative fashion which remains a product of *linguistic* necessity – of the causal link between name and referent – rather than a process of knowledge-acquisition whose stages can be traced through the finding-out of hitherto unknown truths about (say) the molecular composition of water or the microstructural properties that assign the element 'gold' to its rightful place (atomic number 79) in the periodic table of elements. In other words there is still a conventionalist aspect to the causal theory of reference, a sense in which it has to do with the meaning standardly attributed to certain terms in virtue of linguistic precedent or custom. And this despite the claim that it offers an alternative – a causal-explanatory alternative – to those forms of more overtly conventionalist or descriptivist theory that would relativize truth to some particular language-game, discourse, paradigm, ontology, conceptual scheme, or whatever. For such a claim could be justified only if it took adequate account of those extralinguistic (real-world) explanatory factors that alone provide an adequate – non-circular – grounding for the assignment of determinate truth-values through a causal theory of reference.

Thus the trouble with the Kripke/early Putnam account – despite its many advantages as compared with the standard descriptivist model – is that it offers little help in understanding the more complex interactions between thought, language, and reality. That is to say, it narrows the scope of causal explanations to the point where they work only for items (such as 'gold' and 'water') whose reference has long been 'fixed' in the sense that more recent discoveries concerning their molecular or atomic structure have not raised significant problems in respect of their ontological status.

Hence the intuitive self-evidence of Kripke's claim that certain propositions about them – like 'water = H_2O' or 'gold is the metallic element with atomic number 79' – have the character of *a posteriori* necessary truths. What is not so easily explained on this theory is the type of case that Brown examines in his survey of scientific thought-experiments as a source of *a priori* demonstrative proof for conjectures which can then (perhaps with the advent of more refined experimental techniques) be borne out through assisted observation as a matter of empirical fact. Brown offers many such examples. They range from Galileo's thought-experiment with the falling canon-ball and musket-ball to the curious case of Maxwell's demon – conjured up to explain how entropy (or the second law of classical thermodynamics) might conceivably be reversed – and Bohr's calculation of the energy-levels required to 'bounce' a free electron from one to another orbital path when it collides with a second particle. What they all have in common is the twofold premise (1) that we can gain knowledge of the physical world through rigorously-conducted thought experiments, and (2) that such experiments depend for their validity on the truth of ontological (and causal) realism as a necessary working assumption.

Brown's main point is that the current realist *versus* anti-realist debate has been skewed by the legacy of logical empiricism and the flat choice which it appears to offer between two equally unsatisfactory extremes. This tendency is most pronounced in philosophical responses to the various interpretative problems thrown up by quantum mechanics. Thus 'realists are tempted to say that reality is one way or another and that superpositions [e.g., the wave/particle dualism or the impossibility of making a simultaneous measurement of a particle's position and momentum] merely reflect our ignorance'. Anti-realists, conversely, deny this: 'they hold that the micro-world is indeterminate until measurement puts the world into one of the base states'.[17] On the one interpretation we just don't know – and may indeed have no possible means of knowing – why quantum mechanics should in fact have worked so well as a matter of theory and applied technology (transistors, lasers, etc.). On the other, we are quickly into the realm – *via* Schrödinger's cat and suchlike wild hypotheses – of a purely instrumentalist conception of scientific method where truth is whatever we make of it by 'intervening' with this or that purpose in mind.[18] As Popper notes, this argument has some dubious antecedents, from Cardinal Bellarmine's demand that Galileo retract his heliocentric hypothesis for reasons of doctrinal orthodoxy, to Bishop Berkeley's counting reality a world well lost in order to preserve God's indispensable role as omniscient observer.[19] In other words, there is a close (and sometimes an all-too-holy) alliance between instrumentalism, phenomenalism, and idealism, all of which beckon from the further reaches of current anti-realist interpretations in the quantum-theoretical domain.

As a counter-example to such arguments Brown takes the taxonomy of elementary particles developed over the past two decades as the best, most elegant and comprehensive of any model so far proposed. 'The basic assumption', he writes,

> is that matter consists of two types of particles, quarks and leptons, and that there are forces between them which are carried by a third type of particle known as bosons. There are a variety of quarks (up, down, charmed, strange, truth, beauty) which combine in various ways to form the more familiar protons and neutrons. The leptons include the electron, the various neutrinos, the muon, etc. The bosons include the photon (which carries the force for electrical interactions), the vector bosons (which carry the weak force) and gluons (which carry the strong force). Gravitation has resisted all efforts to be successfully incorporated, but it is thought there should be gravitons to carry the gravitational force, as well.[20]

On an anti-realist (or instrumentalist) account this theory would amount to nothing more than an exotic illustration of the way that scientists multiply theoretical entities in order to 'describe' or 'explain' what is actually a construct out of their own current linguistic or conceptual scheme. In support of this case they could point to the sheer proliferation of neologisms – 'quark' first appeared in Joyce's *Finnegans Wake* – and the extreme unlikelihood that the theory as it stands will turn out to have got things right once and for all, and thus to require no further additions to the present array. From which they conclude that any 'truth' the theory may possess is a truth strictly relative to its own descriptive conventions (or instrumental purposes) and in no sense closer to 'reality' than other theories that had previously held the field.

Brown's response to this argument is one that I should have liked to quote at length since it provides a very detailed and convincing case for a realist approach to these issues in philosophy of science. However its gist may be summarized as follows. The quarks and leptons 'come in families', each pertaining to a certain energy-level and with their masses increasing in regular though complex proportion from one 'family' to the next. This order of increase is expressed in units given by a known formula (MeV/c^2) whose value in any given case can be determined both arithmetically, that is, as a matter of *a priori* deduction, and – at least up to certain practical limits – through experiment with high-energy particle accelerators. Thus 'the inclination to assume ever more families of particles is extremely natural' since it is supported not only by striking regularities of a mathematico-deductive kind but also by the best observational evidence to hand. As Brown puts it:

> [w]hen we look at the table of particle families, it seems perfectly obvious that the only thing standing in the way of finding heavier quarks and leptons is the lack of energy . . . . [Yet] the classification scheme does not stop there. It goes on indefinitely. The taxonomy, of course, embodies laws of nature. Thus, 'the mass of the u quark is 5 MeV/c^2' is an example of a law. Just as the taxonomy goes on for ever, so do the laws about ever heavier quarks and leptons, even though they are never instantiated.[21]

To this the convinced anti-realist might respond that the entire theoretical construction rests upon terms (such as 'quark' and 'lepton') which start out as purely arbitrary products of linguistic definition, and whose supposed reality is no more proven than the 'laws' that govern their putative mass in regions – or at levels of energy – beyond the scope of physical attainment. Indeed, so it is argued, the very idea of 'laws of nature' is a fiction adopted for purely instrumental purposes and one that can drop right out of the scientific picture without the least effect upon science's ability to carry on producing the goods. For it does so *not* by getting things right in some ultimate sense – by 'cutting nature at the joints' in Plato's gruesome metaphor – but simply in virtue of its happening to fit with some current set of aims and objectives. In which case this talk about 'quarks' and 'leptons' has no more truth, ontologically speaking, than earlier talk about those various elementary particles – from the atoms of Democritus to the models devised by physicists from Dalton to Bohr – whose 'reality' was likewise a matter of convenient hypothesis.[22]

It seems to me that Brown is entirely justified in regarding this as an extravagantly counter-intuitive and indeed a downright implausible idea. To say that Democritus, Dalton and Bohr were talking about 'different things' is to make nothing more than the trivial point that they brought very different theoretical constructions to bear upon a postulated object – the atom – whose existence Democritus managed to guess at through sheer ungrounded *a priori* intuition whereas later (post-Daltonian) scientists were able both to *verify* its existence and to *specify* its microstructural attributes in far greater depth and detail. Nor is the case for ontological relativity borne out by the fact that new models are constantly proposed and that physicists continue to disagree over just which model provides the most elegant, complete, or satisfactory solution. For it is still self-evident that they must be disagreeing about *something* – that their rival theories must have some shared focus or common point of reference – if such debates are to make any sense. And this applies just as much to those other set-piece examples from the history of science (terms like 'mass', 'gravity', 'acceleration', etc.) that are often cited by followers of Kuhn as establishing the case for radical meaning-change between paradigms, and hence – so it is thought – for 'ontological relativity' as a simply inescapable upshot.[23] For here also there is an unwarranted inference from the fact that *certain terms* can be shown to have undergone a shift of meaning to the idea that no 'translation' is possible from one to another paradigm since those terms dictate the *entire frame of reference* (including the postulated real-world objects) outside of which they become ontologically and epistemologically void.[24]

'What does seem to be central', as Brown rather wearily remarks, 'is the deep anti-realist intuition (that some people just seem to be born with) that the only idea of truth we can have is one that is linked to how we actually determine what is true.'[25] And he then goes on to cite the later (1981) Putnam to the effect that ' "true" in any other sense is inaccessible to us and inconceivable by us'.[26] But the mere fact of some people's being thus temperamentally disposed is of course no argument for the validity of their case, any more than it would be with flat-earth believers, creationists, or latter-day exponents of the phlogiston-theory. Indeed one can find Putnam urging precisely the opposite case – and with equal conviction – in a passage from one of his earlier books. Thus:

The positive argument for realism is that it is the only philosophy that doesn't make the success of science a miracle. That terms in mature theories typically refer . . . , that the theories accepted in a mature science are typically approximately true, that the same term can refer to the same thing even when it occurs in different theories – these statements are viewed by the scientific realist not as necessary truths but as part of the only scientific explanation of the success of science, and hence as part of any adequate description of science and its relation to its objects.[27]

Of course it may be said – and quite rightly – that the issue cannot be decided either way by merely quoting Putnam against himself and thus showing (what is in any case widely known) that he has changed his mind on this particular topic. Still one is not obliged to go along with his later position to the point of conceding that any truth in such matters is 'internal' to some given frame of reference or context of belief, and hence that there is *ultimately* no deciding between the arguments advanced by Putnam 1 and Putnam 2. For this would of course amount to a verdict in favour of Putnam 2, the anti-realist (and relativist to all intents and purposes) who now thinks his earlier position simply untenable.

There is a parallel here with Rorty's occasional attempts to appear even-handed as between, say, realist and anti-realist approaches, or 'constructive' and 'edifying' ways of thought, while in fact this very gesture – treating any choice as simply a matter of personal or cultural preference – is such as to suspend all questions of truth or criteria of valid argument, and hence to endorse item two in each pair.[28] This rhetorical trick has its obvious uses in maintaining a semblance of equitable treatment for any viewpoint – scientific realism included – which some thinkers may continue to find 'good in the way of belief'. But it also enables Rorty to reserve certain handy pragmatist options, among them the idea that while science can tell us nothing 'true' about a 'real' world it can none the less foster desirable attitudes within and beyond the scientific community. 'On this view', he writes,

there is no reason to praise scientists for being more 'objective' or 'logical' or 'methodological' or 'devoted to truth' than other people. But there is plenty of reason to praise the institutions they have developed and within which they work, and to use these as models for the rest of culture. For these institutions give concreteness and detail to the idea of 'unforced agreement'. Reference to such institutions fleshes out the idea of a 'free and open encounter'.[29]

This passage may remind us of those fallback responses to logical empiricism which conceded the validity of that programme for all matters of scientific, factual, or logically accountable truth, but which sought some refuge for human values – for ethics, religion, and poetry – in a separate realm of 'emotive' or attitudinal response.[30] Rorty gives the argument a different, supposedly more democratic spin by softening the science/culture dichotomy (since science is no longer to be thought of as epistemically privileged), and locating those values at a point of convergence between the private-individual and the social or collective spheres. Hence his vaguely Habermasian appeal to science – or scientific 'institutions' – as just the sort of thing we should wish to emulate in the democratic quest for 'unforced agreement' through 'free and open encounter'.

However it is hard to conceive how scientists or anyone else could maintain a respect for such values by adopting the outlook which Rorty recommends. What this amounts to is a version of the 'as-if' attitude, the idea that certain beliefs are good – conducive to moral, social, and political well-being – irrespective of their truth-content or quite apart from any question concerning their validity as a matter of reasoned or principled argument. In other words one is required to accept them as 'true' for the sake of their (presumed) psychological or wider humanitarian value while at the same time knowing – perhaps in some region of the left cerebral hemisphere – that this truth is merely a convenient fiction adopted with such felicific ends in view.[31]

To point out that this simply won't work as a matter of applied psychology is a weak response in the obvious sense that it accepts Rorty's own pragmatist definition of what is 'good in the way of belief'. The argument has to start much further back, at the stage where anti-realism first gets a hold through that now-familar pattern of retreat from the tenets of a narrowly positivist conception of language, logic, and truth. That Rorty is still in the grip of this conception – or of the problems to which it gave rise – is evident in his constant rehearsal of the case not only against logical empiricism and its various derivatives but also against any truth-based or realist approach that displays the least concern with these and kindred questions. What he thus fails to recognise is the sheer implausibility of a theory which – as Putnam once described it – renders the success of science a 'miracle' by denying that terms in mature scientific theories 'typically refer'; by declaring those terms to be non-translatable (or semantically opaque) from one such theory to the next; and by rejecting any notions of 'reality' or 'truth' save as optative constructs out of this or that language-game, paradigm, preferred 'vocabulary', or whatever.[32] At the limit one may have to concede, again like early Putnam, that some people – the later Putnam included – are temperamentally disposed to endorse such claims and just *cannot see the point* of any realist counter-argument from the conditions of possibility for the scientific enterprise and for knowledge-acquisition in general. But this is no reason to accept their case as valid by the best (philosophically accountable) standards of reasoned enquiry or even as making adequate sense on their own preferential terms. For it can easily be shown – as Brown points out – that the later Putnam cannot have it both ways, criticizing Rorty for his relativist stance while himself opting for a theory (that of 'internal realism') which amounts to pretty much the same thing under a slightly less provocative description.

Thus in 1984 – well into his later phase – Putnam can still be found rehearsing all the standard objections against relativism, and in particular against Rorty's idea that 'rightness is simply a matter of what one's "cultural peers" would agree to, or worse, that it is defined "by the standards of one's own culture" '.[33] More than that: he can still praise Kuhn for having come to recognize – in a marked shift from his previous position – that 'rationality and justification are presupposed by the activity of criticizing and inventing paradigms and are not themselves defined by any single paradigm'.[34] But the question remains as to just what those standards of rationality and justification can be exercised upon – or directed toward – if one adopts the late-Putnam compromise view that reality and truth are somehow 'internal' to the particular descriptive framework (for which read 'language-game', 'paradigm', 'discourse', etc.) that decides what shall count as a veridical statement in any given case. As Brown more trenchantly puts it: 'How does Putnam himself avoid falling into the very relativism he rightly despises? With "independent reality" banished, how is it possible to maintain the distinction that objectivity demands, the distinction between "being right" and merely "thinking we are right"?'[35]

From Rorty's point of view such talk of 'objectivity' is hopelessly beside the point. It is just another variant of the old *quid juris* question – the argument from so-called 'conditions of possibility' – on which philosophers from Plato to Kant and Husserl (not to mention their present-day analytic *confrères*) have wasted such a deal of time and intellectual energy.[36] Putnam cannot quite bring himself to adopt this breezily dismissive line, knowing as he does that the achievements of science – and our knowledge of the growth of scientific knowledge – are such as cannot possibly be explained by a Rortian appeal to the shifting conventions (or the periodic switches of vocabulary) that constitute 'progress' as viewed from within our own cultural neck of the woods. As Brown remarks, '[t]his is as true of the new Putnam, the anti-realist, as it was of the old Putnam who used "truth" to explain the success of science'.[37] However, it is far from clear that he can have any argument – or any justification consistent with his own later view – for this claim to hold the line between a relativism that reduces to self-contradictory nonsense and an 'internal realism' capable of handling the standard anti-relativist objections. For one can' well imagine Rorty coming back with the argument that there is just no difference (in pragmatist terms: no difference that makes any difference) between conceiving truth as 'internal to' or as 'relative to' some given language-game, conceptual scheme, frame of reference, and so on. Thus Putnam's case against Rorty is one that applies with equal force to his own position as adopted in his writings from the mid-1980s on.

WORLDS WITHOUT END: NELSON GOODMAN

This issue about realism is of course closely bound up with the issue about explanatory laws in the physical sciences and their status *vis-à-vis* the conduct of mature or progressive scientific enquiry. Logical empiricism is the main inheritor of Hume's deep scepticism in this regard, his belief that philosophy could produce no evidence (at any rate no logically compelling proof) that causal explanations were anything more than a version of the *post hoc, propter hoc* fallacy, that is, our indurate habit of supposing some necessary connection between events where in fact all we witness is 'regular succession' or one thing happening after another.[38] Brown's counter-arguments are especially revealing, as I have said, because they focus on the role of thought-experiments as a test-case for the anti-realist (or the radical empiricist) claim that no putative 'laws of thought' can possibly have any bearing on this matter of *de re* as opposed to *de dicto* necessity. What we learn from such experiments – Galileo on falling bodies, Maxwell's demon, Einstein's elevator, Bohr on the quantum-mechanical laws of mass-energy conversion – is the fundamental error of thinking, like Hume, that there cannot in principle exist any means (any genuine-ly causal-explanatory means) of bridging that gulf between the orders of material and logico-conceptual necessity. For this scepti-cal view is itself the product of a narrow and impoverished empi-ricist outlook joined to an equally restrictive idea of what constitutes logical enquiry. The limits of both have been amply shown up over the past four decades of intensive criticism directed at the logical-empiricist paradigm and its various successor doc-trines.[39] But this has not been enough to save many thinkers – Rorty and the later Putnam among them – from embracing just the sorts of sceptical conclusion to which that movement very soon gave rise.

For Brown, on the contrary, 'the advantages of a realist view of laws are immediately apparent'. Thus:

> [t]o start with, this account distinguishes – objectively – between genuine laws of nature and accidental generalizations. Second, laws are independent of us – they existed before we did and there is not a whiff of relativism about them. Thus, they can be used to *explain* and not merely to describe events. But most important, notice the justice it does to our intuitive understanding of the standard model. Laws on the platonic view are not parasitic on existing objects and events. They have a life of their own. Even if the universe should not have enough energy within it to produce very heavy quarks and leptons, there can still be laws about such things.[40]

Hence the significance of thought-experiments as a means – some-times the only means – of extending knowledge into regions beyond the current limits of empirical research or physically-attain-able 'reality'. In such cases there is no substitute for the kind of rigorously-framed hypothesis that allows theories to be tested and results arrived at on the basis of a strong *a priori* claim concerning what must necessarily hold – as a matter of invariant law – across and despite the distinction between real-world and possible-world domains.[41]

For this purpose the 'possible worlds' in question must of course be construed as *physically* possible – that is, as compatible with all the laws of physics to the best of our current knowledge – rather than as logically possible through some stretch of counterfactual positing. But one can still make the point (against Quine and other exponents of ontological relativity) that counterfactual hypotheses themselves fall into two main categories. On the one hand, there are those that lack any present real-world instantiation but which contravene no law – no established principle of physics – regarding the range of possible objects, processes, or events whose existence or occurrence is fit matter for scientific thought-experiments. On the other are those – such as phlogiston, centaurs, or Homer's gods – whose existence-claims we have every reason to reject not only on empirical or evidential *but also on counterfactual grounds.*[42] For it is precisely the role of counterfactual hypotheses in the scientific domain to rule out certain false premises (since they generate results that conflict with existing knowledge or evidence) or conversely to establish certain necessary truths (since their contrapositive would lead to some insupportable outcome or manifest absurdity).[43] In short, such hypotheses operate on the same basis – with the same truth-conditions or criteria of valid application – that Brown deduces from his detailed account of those various thought-experiments which have served to advance the state of scientific knowledge. Which is also to say that their probative character is partly but not entirely a matter of their logical consistency with premises whose truth must be accepted *a priori* in order for such reasoning to gain any purchase. What is further required is that the hypothesis or the thought-experiment in question be taken as referring to some real-world (known or discoverable) object domain such that any findings arrived at will possess causal-explanatory as well as logically compelling force. In Brown's words: 'there is more to a law of nature than a regularity or a description of a set of occurrent facts, and that something more is not captured either by subjective attitudes or by ideal deductive systematizations – that something extra must be in reality itself'.[44]

Much has been written about the history of scepticism with regard to such putative laws of nature, beginning with Hume and continuing in our own day with the ontological-relativist arguments of philosophers like Quine and Nelson Goodman.[45] On

Goodman's account there is no way out of the dilemmas be-
queathed by logical empiricism except by adopting a radically
conventionalist approach, one that relinquishes the vain quest for
any privileged scheme – any correspondence-theory or suchlike –
that would link up sentences, *via* beliefs, with causal regularities or
real-world states of affairs.[46] In his case one can see very clearly
how the sceptic disposition takes hold through the use of counterfactual hypotheses (or thought-experiments) conducted solely with
a view to their speculative outcome and without regard to any
realist constraints upon their range of object-terms and predicates.
Hence Goodman's famous 'new puzzle' of induction, his ingenious
restatement of the Humean dilemma arrived at by inventing the
colour-term 'grue', taken to apply to all green objects – say em-
eralds – before a given date and to all blue objects thereafter.[47] (Of
course just the opposite would hold for its symmetrical counterpart
'bleen'.) Granted such examples have a role to play – a well-defined
heuristic or investigative role – in those areas of modal logic where
the relevant 'possible worlds' (and the issue of epistemic 'access'
between them) can be treated in purely abstract terms as a matter
of stipulative warrant. But the case is very different when this
principle is extended, as in Goodman's book *Ways of Worldmaking*,
to the point where it entails a doctrine of strict *ontological* parity for
every such descriptive framework or choice among the various
conceivable schemes for distributing terms and predicates.[48] For it
then gives rise not only to paradoxical (counter-intuitive) results
but to findings that decree an attitude of blanket scepticism toward
even the best-established items of scientific knowledge and the
methods of enquiry that produced them.

   This is why Goodman's ultra-conventionalist stance has provoked a dissenting – at times mildly scandalized – response from
philosophers (like Quine and Joseph Margolis) who are otherwise
committed to a strong version of the ontological-relativist case.[49]
What his argument shows very clearly is the way that this doctrine
gets into conflict with any general claim – such as Quine would
certainly wish to endorse – for the privileged status of the physical
sciences as a model for our best current notions of truth and
method. That is to say, it demonstrates the difficulties that arise in
preserving even a moderate degree of realism (sufficient to account
for the manifest successes of the scientific project) if one denies the
existence of any necessary link between the order of *de re* causal
explanations and the order of inductive or hypothetico-deductive

reasoning whereby those explanations are arrived at. Thus for Goodman, quite simply, 'there is ... no such thing as the real world, no unique, ready-made absolute reality apart from and independent of all versions and visions'.[50] No doubt this is true – trivially so – when interpreted in *epistemological* terms, that is, as a statement of the Kantian point that we can have no access to reality except by way of those concepts and categories that organize the manifold of phenomenal experience. It is likewise perfectly valid when extended to philosophy of science and the various instrumental technologies (electron microscopes, cloud-chambers, particle accelerators and so forth) which make it possible for enquiry to pass beyond the limits of unaided human observation. In this sense one can readily agree that there is no 'real world', no 'unique, ready-made absolute reality' that would somehow be accessible to human knowers quite apart from all their means of knowledge-acquisition. However Goodman is not content to let the case rest at that. For there are, he continues, 'many right world-versions, some of them irreconcilable with others; and thus there are many worlds if any. A version is not so much made right by a world as a world is made by a right version'.[51]

It is at this point – with his turn toward an all-out conventionalist or constructivist theory – that Goodman's case becomes both philosophically incoherent and an affront to every standard of reasoned enquiry in the natural and human sciences. For there is no way of construing the above passage – and many others to similar effect – except as entailing the absurd consequence that reality *just is* whatever we make of it under this or that set of arbitrarily chosen descriptive conventions. We can project as many diverse (even 'irreconcilable') worlds as turn out to be consistent with our preferred methods of projection or our idea of what counts as a valid theory, ontological scheme, inductive procedure, or whatever. And since these are *ex hypothese* just as many and various – viz. Goodman's 'new riddle' of induction – we had better accept that all such worlds (however logically, metaphysically, historically, or scientifically remote from our own) have an equal claim to exist and to function according to laws that are none the less 'real' for our having invented or 'projected' them. 'If we make worlds', Goodman writes, 'the meaning of truth lies not in these worlds but in ourselves – or better, in our versions and what we do with them'.[52] All the same, as Margolis perplexedly remarks, he still wishes to hang on to some Peircean notion of truth-at-the-end-of-enquiry, some idea that – in Goodman's words – 'right versions are different from wrong versions'. And this despite his just having roundly asserted that 'there is no version-independent feature' of any such projected world since 'everything including individuals is an artefact'.

Margolis's comment is worth quoting here since it points not only to a manifest *non sequitur* in Goodman's radically conventionalist view but also to problems with his own and other variants of the more moderate ontological-relativist case. Thus:

> if *we* ourselves are not artefactually made within particular world-versions but *make* them instead, then it is difficult to see how Goodman can escape (some part of) the cognitive privilege he denies; and, if we *are* artefactually made within them, then it is difficult to grasp what it would mean to claim a distinction between right and wrong world-versions. There is no evidence that Goodman ever addresses the dilemma, or puts it to rest, or outflanks its apparent threat.[53]

In other words, there is no reconciling Goodman's extreme conventionalism and anti-realism – his case for the version-dependent nature of *reality itself* and not just our knowledge of reality – with his idea that worlds (or world-versions) can be somehow ranked on a scale of relative 'rightness' and 'wrongness'. For this latter requires that the worlds in question be *not* just 'projectable' according to some arbitrary range of descriptive conventions. Rather they must be capable of assessment in terms that have to do with both the particular world-features concerned and our means of cognizing them as agents who stand in a particular (knowledge-constitutive) relation to just those salient features.

In his criticism of Goodman – I would suggest – Margolis is on the very verge of endorsing a realist ontology and a causal-explanatory epistemology that between them offer the only adequate solution to these sceptical-relativist dilemmas. That he cannot quite bring himself to accept this solution (and indeed very promptly veers away from it) is understandable given Margolis's own strong relativist leanings and his belief that philosophy has been led astray through its over-reliance on a false idea of 'scientific' objectivity and truth.[54] Still he finds Goodman most vulnerable when the question is raised as to 'what makes a category right' – right, that is to say, for descriptive or explanatory purposes – aside from its merely happening to fit with this or that favoured (but wholly conventional) projection-scheme. To which Goodman responds: '[v]ery briefly, and oversimply, its adoption in inductive practice, its entrenchment, resulting from inertia modified by invention'.[55] Thus inductive validity, on Goodman's view, provides an example of 'rightness other than truth'. That is, one can take inductive procedures – in the natural sciences and elsewhere – as perfectly valid by their own practice-specific criteria of 'rightness', and thus head off the usual sorts of knockdown anti-relativist argument, while making no appeal to any 'absolute' (or practice-transcendent) criterion of truth. But here again Goodman gets into a fix when trying to explain how these context-relativized standards of 'rightness' can possibly do without that further criterion or ground of appeal if they are not to invite all the same counter-arguments. As Margolis puts it (with interpolated passages from Goodman): 'although "we cannot equate truth with acceptability" (since "we take truth to be constant while acceptability is transient"), nevertheless *ultimate* acceptability – acceptability that is not subsequently lost – is of course as steadfast as truth'.[56]

So Goodman is back with the Peircean conception which he thinks can be kept on the side of rightness (rather than truth) since it treats the latter as a regulative notion whose virtue is to promote continued enquiry despite the lack of 'absolute' criteria. But this argument simply won't work if one supposes – like Goodman – that 'rightness' must always be relativized to some given descriptive framework or projective scheme which can only be one among the many alternative (perhaps 'irreconcilable') world-versions on offer. As Margolis puts it:

the rather Peircean long run ('entrenchment') that Goodman appears to count on would *require one* actual world with respect to which our versions are fallibilistically and progressively linked, would *require* a world that is not *merely* an artefact (of our conceptual world-versions), would *require* a universal community of enquiring minds functioning self-correctively *with respect to* the one world to which they belong.[57]

One could wish for no better, more incisive statement of the problems with Goodman's attempt to make sense – philosophically adequate sense – of his own ultra-relativist theses. However it would be wrong to interpret Margolis as in any way *endorsing* the sorts of

alternative ('one world', truth-based and realist) view to which he
finds Goodman willy-nilly committed whatever his efforts to avoid
them. As I have said, his response – like Quine's – is more in the
nature of a puzzled enquiry as to why Goodman should have put
the relativist case in so extreme and indefensible a form, and thus
provided extra ammunition for those (the majority of philosophers
from Plato down) who have attacked relativism as ethically bankrupt and conceptually incoherent. Thus Margolis lays out the above
set of claims *not* by way of a proffered solution to Goodman's
self-inflicted dilemmas but as a cautionary instance of the perilous
ease with which a weak (overstated though underargued) relativism can be shown to conceal just such a range of unwitting realist
and truth-based premises.

In fact Margolis's book (the final volume of a trilogy) is devoted
expressly to redeeming relativism from that long history of contemptuous dismissals and developing a version of it which – unlike
Goodman's – yields no hostages to fortune in the way of naive or
unguarded formulations.[58] One motive for this ambitious under-
taking is his belief that much recent philosophy (from logical empiricism on) has been driven into various needless dilemmas by a
misplaced reverence for the methods and truth-claims of the natural sciences. Thus he looks outside the analytical tradition, narrowly construed, to those alternative 'continental' sources – Heidegger,
Gadamer, Ricoeur, and (more reservedly) Derrida – for some poin-
ters toward a more genial conception that would treat science as
just one form of humanly meaningful activity, a project whose
significance can only be grasped through the kind of jointly narrative and depth-hermeneutic approach that renounces such false
scientistic ideals. On Margolis's account these latter were epitomized in the 'unity of science' movement which gave pride of
place to physics and placed other disciplines in a ranking-order
from chemistry to biology (or the life-sciences) and thence to philosophy, anthropology, psychology, sociology, and poor-cousin as-
pirants like literary criticism. To which he responds by arguing, on
the contrary, that we can interpret the truth-claims of the natural
sciences (physics included) only by taking prior account of those
horizons of intelligibility – those narrative or hermeneutic contexts
– that render them humanly knowable. In other words the order
of priority is exactly reversed and philosophers of science are
best advised to seek inspiration in the humanistic disciplines –
literary criticism prominent among them – rather than continue

their fruitless quest for an 'objective' (physics-based) method and ontology.[59] The problem with this emerges most clearly when Margolis reproaches Hans-Georg Gadamer for having recognized the historical contingency of all our truth-claims and yet – at crucial points – retreated into a 'conservative' position that holds out against the perceived threat of a thoroughgoing cultural relativism. 'With one voice', Margolis writes, 'he [Gadamer] coherently affirms that "The historical movement of human life consists in the fact that it is never utterly bound to any one standpoint, and hence can never have a truly closed horizon."' But with another voice he makes statements like the following:

> When our historical consciousness places itself within historical horizons, this does not entail passing into alien worlds unconnected in any way with our own, but together they constitute the one great horizon that moves from within and, beyond the frontiers of the present, embraces the historical depths of our self-consciousness. It is, in fact, a single horizon that embraces everything contained in historical consciousness.[60]

I agree with Margolis that such passages confuse the issue; that nothing is gained by portentous talk of the 'one great horizon' encircling our efforts to communicate across real distances of cultural, linguistic, and historical context; and moreover that Gadamer's position is deeply 'conservative' in so far as it involves this Heideggerian appeal to an order of authentic historical awareness transcending all mere particularities of time and place. So far his complaint seems fully justified: that 'Gadamer is motivated to escape the sinister possibilities of a relativism that may border on complete anarchy; but his conceptual maneuver is entirely unnecessary and remains an intellectual scandal.'[61] On the other hand, it seems to me that Margolis draws just the wrong conclusion from Gadamer's unfortunate plight. That is to say, he regards it as 'unnecessary' – even a 'scandal' – because relativism is not, as Gadamer supposes, a threat to all shared human values or standards of civilized enquiry but a perfectly valid option which Gadamer himself (in more sensible moments) appears quite willing to embrace. Thus he figures, along with Nelson Goodman, as a thinker of albeit very different philosophical persuasion whose work becomes 'scandalous' at just that point where it fails to take the relativist lesson thoroughly to heart.

Nevertheless Gadamer is useful, so Margolis argues, if we discount these more dubious claims and hang on to his idea of historico-cultural-linguistic 'horizons' as the context of all meaningful enquiry in the natural and human sciences alike. Thus: 'the theory (quite apart from Gadamer's use of it) does suggest how to reconcile (1) the realism of cultural phenomena; (2) their ontic indeterminacies; (3) methodological rigour regarding pertinent truth-claims; and (4) what Gadamer is at pains to oppose – the vindication, both ontologically and methodologically, of a moderate form of relativism'.[62] But there are problems here, as likewise with Margolis's attempted retrieval (or rescue-operation) when confronted with the 'scandal' of Goodman's *Ways of Worldmaking*. That is to say, it is far from clear that Gadamer's hermeneutical commitments can be thus neatly separated out into those that suit the 'moderate relativist' case and those that are best passed over in tactful silence. For it is precisely where difficulties arise in making sense of that case – in explaining, for instance, how communication is possible across and despite differences of cultural 'horizon' – that relativists are often driven to adopt some version of the Heideggerian appeal to depth-ontological (history-transcendent) grounds of shared understanding.[63]

Margolis himself, in the above-cited passage, seems to recognize just this problem when he sets out the various alternative conditions for a 'moderate' relativism subject to no such disabling liabilities. Thus there is no making sense of his first desideratum – 'the *realism* of cultural phenomena' – if one takes it (like the later Putnam) that this claim should properly be construed 'realism relative to this or that paradigm, language-game, conceptual framework, etc.'. Furthermore (2), it is difficult to see what could *count* as an example of 'ontic indeterminacy' in the absence of a real-world ontic domain whose normally determinate (objective, invariant, or mind-independent) features are such as make it possible for investigators sometimes – as in the case of certain quantum-mechanical phenomena – to define precisely the limits placed upon our powers of description or measurement. Margolis himself comes close to acknowledging this point when he concedes (item 3) that any tenable or 'moderate' version of the relativist case will have to make room for 'methodological rigour regarding pertinent truth claims'.

However, if it is then asked: ' "pertinent" in just what respect?' there are only two possible answers. One – following Goodman – would take the line: ' "pertinent" just in so far as they accord with this or that descriptive or projective framework'. This answer is plainly inadequate for all the reasons that Margolis has shown. But in that case the pertinence of such claims – as well as the 'methodological rigour' required to establish them – can be 'relative' only in the sense that Bertrand Russell acknowledged himself a relativist. That is, they are true not 'relative to a framework' but 'relative to what is the case'.[64] Of course each claim will need to be assessed with reference to the kinds of methodological rigour that obtain within various fields of knowledge or disciplines of enquiry. Thus the validity-conditions will differ widely as between (say) those that apply in the various natural-scientific domains and those that have developed – through a long-term process of critical and methodological reflection – in fields such as history, sociology, anthropology, and the other human sciences. But there is still something deeply amiss with Margolis's idea that we should henceforth invert the received (science-dominated) order of priority and treat *all* truth-claims as embedded in a cultural or 'narrative' context which renders them opaque to any but a depth-hermeneutical mode of understanding. For there can then be no way of maintaining those criteria – of 'pertinence' and 'methodological rigour' – that Margolis quite explicitly wishes and needs to maintain if his argument is not to collapse into a relativism of the wholesale (Goodman) variety.

6

Stones and Pendulums: Joseph Margolis and the Truth about Relativism

A CASE FOR PROTAGORAS

As we have seen, Margolis considers it something of a scandal that relativism has received such a bad press from philosophers down through the ages. The odds have been stacked against it, he argues, ever since Plato set things up so that Socrates could run rings around Protagoras by misrepresenting his position.[1] And it is still pretty much taken for granted – at least by professional philosophers – that any statement of the relativist case will either turn out to be self-refuting for reasons first explained by Plato and Aristotle, or else have resort to some alternative (for example, pragmatist or instrumentalist) variant which avoids defeat only by adopting a more moderate fallback line.[2] Of course there are some, like Nelson Goodman, who throw caution to the winds and see no need for any such elaborate defences. Thus for Goodman, quite simply, 'there is . . . no such thing as the real world . . . apart from and independent of all versions and visions'. Rather, 'there are many right world-versions, some of them irreconcilable with others; and thus there are many worlds if any'. And lest any doubt remain: 'A version is not so much made right by a world as a world is made by a right version.'[3] In which case there exist as many worlds (or 'realities') as there exist world-versions, descriptive frameworks, language-games, conceptual schemes or whatever, all of them ontologically on a par and none laying claim to a privileged status in point of actuality or epistemic access.

Margolis is frankly embarrassed by Goodman's baroque proliferation of worlds, as likewise by his breezy dismissal of the standard anti-relativist arguments.[4] It is a feeling shared by other philosophers, Quine among them, who have themselves done much to promote the idea of ontological relativity but who regard Goodman as having brought it into disrepute by this reckless manner of proceeding.[5] The case can be presented much better, Margolis thinks, by engaging the opponents on their own ground and coming up with arguments for a 'robust' relativism which won't yield so many hostages to fortune by way of loose formulations and begged or suppressed premises. This position receives its most spirited defence in his recent book *The Truth About Relativism*.[6] Here Margolis sets out to rescue Protagoras from the enormous condescension of posterity and to render his arguments in a form maximally proof against present-day jibes and rebuttals. However, my attention will be focused mainly on some chapters from his earlier volume *Texts without Referents* (1989), the last instalment in a trilogy that lays fair claim to constitute the most sustained, ambitious and resourceful defence of relativism in recent times.[7] Without doubt Margolis has read more widely – in epistemology, philosophy of science, philosophical semantics, aesthetics, hermeneutics, historiography, social and political theory – than could well be expected of any philosopher in a single working lifetime. He is equally at home with debates in the (so-called) 'analytic' and 'continental' schools, pointing up a range of kindred themes which cut right across that largely artificial or totemic divide. All of which will, I hope, justify my treatment of Margolis's book as an impressive but deeply problematical attempt to situate relativism – properly understood – at the end of the road that so many thinkers are currently travelling.

MARGOLIS AND DANTO ON HISTORICAL UNDERSTANDING

These problems appear most sharply when Margolis addresses the awkward question as to why, on his own account, we should draw any firm ontological distinction between past and future events, or history as that which has already happened (and hence cannot be altered) and futurity as that which offers a range of alternative (yet-to-be-decided) possible outcomes.[8] It is, he remarks, a 'peculiarly treacherous issue' not only for philosophers inclined – like himself – to complicate the terms of that distinction but also for anyone concerned with the uses and abuses of so-called 'revisionist' historiography. Margolis's response is none the less to grasp this particular nettle and deny that there are or could be any adequate ontological grounds for viewing past events as in some way sealed off from present or future revision. 'What is the historical past? Is it alterable, and, if it is, how and in what sense can it be changed?' (*Texts without Referents*, p. 293). According to Margolis this question has met with a 'nearly universal' answer, one which – subject only to 'minor adjustments' – insists that 'if it is real, then the past is fixed and unchangeable'. These adjustments mostly take the form of allowing that '(a) it can acquire new *relationships* with an evolving present, and (b) we can always abandon, for cause, a previous *characterization* taken as historically true, on the basis of new, even hitherto nonexistent, evidence' (ibid.). That is, they locate any changes in our understanding of history very firmly on the side of *our understanding* – for example, our access to new sources of evidence – rather than on the side of history itself.

Margolis sees this as a shuffling or compromise solution which simply fails to meet the relativist challenge. For it is a theory of history that fits well enough with all the main tenets of the unity-of-science programme, namely (as Margolis describes them): (i) the espousal of a 'nonreductive physicalism'; (ii) the requirement that 'causality, whether physical or historical, is and must be nomologically constrained'; (iii) the asumption 'that truth is, in principle, decidable'; (iv) 'the correspondence theory of truth' (in however refined or qualified a form); and (v) 'traditionalism, which repudiates relativistic truth-values, even as it admits the flux of history and refuses the blandishments of transparency claims'. (p. 294) In other words this theory cleaves to the 'traditionalist' – for which read 'positivist' – view of history as a strictly closed ontological domain where events occurred, where actions had consequences, and where truth-values hold (or facts remain facts) despite any changes in our knowledge or interpretation of them. Thus Margolis cites Arthur C. Danto as a leading exponent of the received idea that 'reference makes realists of us all', and moreover that '*historical* reference is simply reference in a certain temporal direction relative to the referring expression itself'.[9]

However this could only be the case, Margolis argues, if historical events stood in the same relation to historical knowledge as that which obtains – on the logical-empiricist account – between observational data and our various statements, propositions, or theories concerning them. In short it assumes what cannot be proved and what Margolis finds frankly incredible: that there exists 'a seamless conceptual connection between physical science and human history' (p. 295). For he can see no other way of construing Danto's qualified defence of a covering-law (deductive-nomological) approach to issues of historical interpretation. As Danto sees it the historian's business is perhaps to start out from an 'internal' perspective, one that attempts to recreate history from inside the minds, the experience or worldview of those who were living at the time. Such was of course Collingwood's prescription in *The Idea of History*, and such the broadly narrative-hermeneutical approach favoured by thinkers like Gadamer and Ricoeur.[10] But it is precisely the mark of a genuine historical *science* that it should then move on to raise 'external' questions – questions of representational accuracy or determinate truth and falsehood – with regard to the various ideological values placed upon events by interpreters who lacked this requisite critical distance.

Danto instances the Paris Commune as an episode that aroused (and has continued to arouse) such intense political feelings as to make it a useful test-case for his distinction between the two orders of historical understanding. Thus:

> It is not what the Paris Commune was, but what it has come to mean to radical and conservative alike which determines the political complexion of the present. . . . But to point out, as good scholars repelled by falsehood, how off the mark [they and their age may] have been relative to the *en-soi* of the times, is in a way to forfeit a proper understanding of their age *from within*. . . . And indeed, we treat them as historians-as-scientists when we raise the question of the truth or falsity of their representations. . . . Representations, historical or any other kind, are within and without reality at once.[11]

Margolis finds nothing here but a failed and self-contradictory attempt to rescue historical knowledge from the bugbear threat of a wholesale scepticism which renders such knowledge utterly impossible. Thus, in Danto's words, '[h]istorical beliefs are [both] internal and external to historical reality, and it was the curious muddle of relativism to have denied the latter by having discovered the former'.[12] The upshot of this is a purely 'internalist' approach that equates the truth of past events with the way they might have struck a contemporary or – just as bad – with their shifting afterlife of ideologically partisan visions and revisions. In which case (as Danto sees it) the issue comes down to a conflict of principle between those who seek a more adequate and truthful understanding of history and those who – following Nietzsche, Foucault, and other proponents of the strong revisionist line – reject such 'positivist' claims out of hand. On this view historical events have no more reality (and impose no greater constraints upon the range of warranted interpretations) than future events or counter-factual hypotheses adopted for whatever speculative purpose. Thus the past can be endlessly altered or rewritten in the interests of a Nietzschean *wirkliche Geschichte*, an approach that renounces the false ideals of science, objectivity, or truth, and which denies history any relevance or value except in so far as it answers to our present needs and desires.[13] Such – according to Danto – is the prospect embraced (knowingly or not) by those who adopt the 'internalist' perspective as a substitute for genuine, disciplined historical enquiry.

Margolis sees this as a panicky over-reaction and a sign that Danto is still in thrall to that false (science-dominated) notion of 'truth' which has wrought such confusion in the wake of logical empiricism. On Danto's account '[t]here is no need, apparently, to sort out ontologically appropriate objects' (*Texts without Referents*, p. 295). Thus history becomes just a sub-genre of scientific knowledge where actions and events must be understood in much the same way – and according to much the same standards of veridical utterance – as apply to objects, observations and theories in the physical domain. It is concerned 'entirely with the description of *anything* viewed with respect to a certain temporalized reference (in, say, the physicist's sense)'. From which it follows – on Margolis's reading of Danto – that 'the theory of historical reference is a purely formal matter, indifferent as such to *what* it is a history about' (ibid.). But this simply won't work as an adequate account of what goes on in the process of historical understanding. For that process must involve something other and more than the reference to past events conceived through the wholly misleading analogy with physical observation-data. Historical data *just don't exist* – are simply not available for inspection – in the way that such theories propose. There are problems enough (so Margolis contends) with this approach as applied to the natural sciences, let alone the historical-interpretative disciplines where nothing could plausibly serve as a stand-in for the 'objects' or 'data' of empirical research.

The account might just be thought credible, Margolis concedes,

> *if* one believed (as Danto apparently does) that the historical
> world behaved with respect to truth in the same way as does the
> physical world, if (say) something like a nonreductive physical-
> ism (or even a reductive physicalism) obtained. It is not merely
> (or not yet at least) that Danto's is a false claim about history. It is
> rather that Danto never lays a proper ground for his own particu-
> lar claim. (*Texts without Referents*, p. 295)

Margolis thinks that Danto cannot and could not possibly 'lay a
ground' for this claim since there *is* no ground – no half-way
convincing or plausible argument – for treating historical knowledge in these terms. The argument would work only on condition
that historical events were objectively cognizable and in some sense
there for adequate description from the standpoint of a neutral
observer. But since they clearly meet neither of these criteria –
belonging as they do (by very definition) to a past that has receded
beyond reach of our present perceptual or cognitive grasp – therefore Danto must be doomed to defeat in his attempt to bring them
under principles derived from the natural-scientific domain. So far
from historical reference 'mak[ing] realists of us all' it serves to
underline the sheer *impossibility* that historical understanding
should ever achieve such 'scientific' status.

However there is more at stake in Margolis's quarrel with Danto
than this long-running issue of whether there exist distinctive
methods or modes of enquiry appropriate to the *Naturwissenschaf-
ten* (natural or physical sciences) on the one hand, and the *Geiste-
swissenschaften* (interpretative or hermeneutic disciplines) on the
other. For it is his contention – pursued single-mindedly through
this otherwise labyrinthine work – that truth-claims *even in the
physical sciences* take rise from a context of humanly significant
meanings, purposes, and intentions outside of which they can
make no sense for later (historically situated) human enquirers.
Hence his particular interest in Danto as a thinker who has got the
relationship upside-down and vainly attempted to reverse that
order of priority. Hence also the connection that Margolis perceives
between realist (especially causal-realist) theories of reference in
philosophy of language and arguments – like Danto's – that purport to establish a science of historical understanding. For it is here
that epistemological debates have a close (though he thinks most

often a dubious) bearing on larger issues of interpretative method and truth.

This connection comes out most plainly in Margolis's treatment of an example that Danto offers in support of his claim for an 'externalist' perspective capable of somehow transcending and correcting the effects of 'internal' (observer-relative) bias. The story has to do with certain stones which – in Danto's fictive but real-world-compatible account – were first put to use during Roman times as weights for market-scales and then found their way into the structure of a Christian church built on the old market site. For Danto this provides a useful, even clinching case in support of the argument that historians *can and must* respect the ontological distinction between objective truth (as witnessed by the stones' continuing material existence) and the various cultural or interpretative contexts that supervene in the course of historical events. As Danto puts it:

> the self-identical stones have come under different and non-over-lapping descriptions for differing sets of peoples who, though they shared these stones, lived in d*ifferent worlds*. . . . To enter another world, as I am now using this expression, would be to see the same objects under different descriptions and against the background of different sets of historical and causal beliefs.[14]

One can see why Margolis italicizes the phrase *different worlds*, recalling as it does Thomas Kuhn's well-known (indeed notorious) claim that the world and its objects or constituent features are quite literally different for scientists working on either side of some major paradigm-shift.[15] Margolis fully approves this claim and signals his acceptance by citing the relevant passages from Kuhn. Thus for instance: '"Lavoisier saw oxygen where Priestley had seen dephlogisticated air and where others had seen nothing at all"; "Lavoisier saw nature differently . . . [he] worked in a different world"; Aristotle saw "only swinging stones" where Galileo saw true pendulums; "pendulums were brought into existence by something very like a paradigm-induced gestalt switch" '.[16]

Nor is Margolis in the least disposed to adopt what some exegetes – the later Kuhn among them – have seen as a possible escape-route from the stance of extreme ontological relativism implied by these and similar passages. That is to say, he rejects the more moderate version on which it is a matter of scientists 'seeing things differently', that is, under different descriptions or explanatory theories, but where the 'things' themselves none the less perdure as real-world objects of enquiry. On the contrary, Margolis has a mildly reproving footnote where he notes Kuhn's later (recidivist) tendency to talk about 'invariant physical laws' that would somehow ensure continuity of reference – and our knowledge of the growth of scientific knowledge – across and despite such paradigm-shifts.[17] 'What, obviously, is called for is a more radical account of the nature of science, physical law, and scientific explanation' (*Texts without Referents*, p. 64n). And by this he means an account that gives up any version of the old ontological-realist belief in a world that actually exists, that possesses certain mind-independent objective features, and of which certain statements hold (or would hold) true quite aside from what we or other observers may happen presently to think concerning it.

So one can see why Margolis remains unconvinced by Danto's strenuous attempt to secure the distinction between history-as-science and history as interpreted from 'inside' the various cultural perspectives brought to bear on it. On his (Margolis's) view the 'different worlds' argument should be pushed all the way so as to encompass not only our ideas, concepts, or perceptions of reality but *reality itself* in its every last physical or ontic determination. This follows, he believes, from the manifest failure of the realist programme – whether in epistemology, philosophical semantics, or philosophy of science – to come up with any adequate arguments in its own defence. Thus Danto's stones are entirely on a par with the mere 'swinging objects' that Aristotle perceived and the stones that Galileo used to demonstrate the laws of pendular motion. That is to say, 'ontological' considerations are quite beside the point, whether as regards the perdurance of real-world objects under differing cultural descriptions, or again with respect to (supposedly) invariant physical 'laws' that undergo the process of scientific paradigm-change. In each case there is nothing – aside from that deep-grained realist prejudice – which should prevent a philosopher of science like Kuhn or a philosopher of history like Danto from acknowledging that these were *not the same stones* when removed from one to another cultural context or interpretative framework.

It is a tempting option, here and elsewhere, to suppose that Margolis cannot be committed to so extreme an ontological-relativist view, or at least that he must be using certain terms ('reality' and 'truth' among them) in a sense that, once recognized, will save his argument from its own more mind-boggling implications. But this option won't work since Margolis goes out of his way to insist – *contra* Danto and the later Kuhn – that the relativist case has to be stated in just this strong (ontological and not merely epistemic) form if it is to stand any chance of meeting the usual objections. Thus he sees no hope of a compromise settlement in Danto's willingness to accept that 'internal' (hermeneutic) approaches to history are valid up to a point, and that they sometimes play an essential – if preliminary – role in our present understanding of past events. For Danto, 'some reference to the *beliefs* of agents [in either the Roman or the Christian world] is required in the explanation, hence the understanding of their [respective] actions [with regard to the stones]'.[18] However Margolis sees this as just another, somewhat desperate holding-operation designed to keep relativism at bay and to prop up the claims of a realist ontology that finds no support from Danto's example. 'Naturally', he writes,

> one is inclined to speak of the history of the *actions and beliefs* (and the like) of the Roman merchants and the Christian believers involved. But what of the history of the *stones*? Also, is it so clear that the stones of 'the one world' *are* the same as those of the other? And what is the meaning of 'same' in this rather complicated context? Danto never supplies an explicit answer to any of these questions. (*Texts without Referents*, p. 296)

Margolis rejects Danto's attempted solution because it still finds room for a realist (or non-reductive physicalist) account of what makes the difference between true and false statements concerning the historical past. That is, it allows just so much ground to the claims of hermeneutic understanding – enough to head off the charge of downright positivist or 'reductive' methodology – while still making sure that those claims are subject to the higher tribunal of objectivity and truth.

For Margolis, on the contrary, any 'truth' to be had is always a truth *internal and relative* to the various changing cultural contexts through which they and we (merchants, believers, present-day historians) can alone gain access to it. And this applies just as much to those material objects – like the stones – that Danto thinks of as somehow bearing witness to a perduring reality beyond or above such shifts of cultural perspective. For this is to adopt a relationist (rather than a thoroughgoing relativist) stance, one that acknowledges the range of possible viewpoints, cultural frameworks, socio-historical contexts of enquiry, and so on, but which none the less assumes that they all must have reference to some real-world ontic or factical domain whose pastness (or non-availability for present inspection) detracts not a jot from its objective, mind-independent status. Margolis thinks that this compromise position is incoherent for all the reasons standardly advanced by anti-relativist thinkers from Plato down. Thus the only alternative for a fully consistent and philosophically defensible relativism is to take this particular bull by the horns and deny that there is *anything* – no matter how stonelike in its seeming brute facticity – to which all the various perpsectives could 'relate' and thus save the day for a realist ontology with minor hermeneutical concessions.

The latter is Danto's line of argument, one that lays him open – so Margolis thinks – to the twofold charge of both defending a wholly untenable realism in respect of historical events and admitting a weak (relationalist) account of cultural contexts and perspectives. Thus 'it is *not* that Danto believes the past is not fixed, but rather that he believes it is open *only* by way of *relational* innovations regarding our beliefs and how we act with respect to the (intrinsically) fixed past' (*Texts without Referents*, p. 297). And again: 'Danto's emphasis on *relational* accretions involving the future (which, *contra* Peirce, do "affect" the past) really do no such thing, since they do not "affect" the inherent or intrinsic properties of "things" ' (ibid.). So the only way out of Danto's dilemma, according to Margolis, is to adopt a really robust (ontological) relativism which affirms that *everything* is a product of interpretation, that is to say, not only our ideas of the past but *the past itself* along with all those actions, events and even objects (physically enduring objects) which Danto mistakenly considers exempt from the vicissitudes of cultural change.

ANTI-REALISM, RELATIVISM, AND THE HERMENEUTIC TURN

I should confess to having read Margolis's book in much the same spirit that he reproachfully attributes to the later Kuhn when attempting to answer his critics by placing a less relativistic construal on the passages that most gave offence. That is to say, I looked hard for any welcome signs that Margolis was *not* after all committed to a form of full-scale ontological relativism whereby the very nature of physical realia – and the very occurrence of past events – would be subject to change or retroactive modification through shifts of cultural perspective. Here and there one finds passages that will just about admit of some such redemptive reading. But they are most often followed (sometimes within the same sentence) by a prompt retraction which leaves no room for manoeuvre on the exegete's part. Thus Margolis speaks at one point of the 'suppressed absence' – the unconsidered alternative – that 'unmistakably betrays itself' in Danto's theory of historical understanding. 'What is suppressed', he continues, 'is simply this: the conceptual option that, *perhaps, the historical past is both real and not fixed in its intrinsic properties* (p. 294; italics in the original). The apparent concession to realism here is a hedge against one obvious line of attack: that his theory reduces the physical world and all its constituent features, past and present, to so many cultural-linguistic artefacts or products of interpretation indifferently ranked with respect to their reality or truth-content. As I have remarked, Margolis is sensitive to such criticism and indeed considers it justified when brought against others (like Goodman) who propound a less circumspect, more wholesale version of the relativist case. But to say that the historical past, although 'real', is 'not fixed in its intrinsic properties' is to lay oneself open to just the same sorts of objection. For it means in effect that interpretation goes all the way down and that 'reality' can be construed only as relative to this or that paradigm, conceptual scheme or interpretative framework.

What Margolis wants – in company with the later Putnam – is a kind of 'internal' realism which would somehow maintain certain plausible truth-conditions (thus avoiding all the standard anti-relativist arguments) while yielding no hostages to fortune in the way of express ontological commitments.[19] But by denying that 'reality' (past or present) has any 'intrinsic properties' – that is, any mind-independent objective features that could serve as the basis for ascribing determinate values of truth or falsehood – Margolis effectively closes off this particular line of defence. And he does so, moreover, by comparing history to precisely that domain of knowledge (i.e., the natural sciences) where there would seem – on the face of it – least justification for supposing that truth might have nothing to do with the 'intrinsic properties' of a mind-independent reality. For in his view those properties are 'real' only to the extent that they have played some role – and acquired certain salient features – throughout the history of changing cultural or interpretative contexts.

Thus according to Margolis it makes no sense to speak of their 'intrinsic' character, attributes, features or causal properties, if by this we mean what the realist standardly means, that is, 'belonging to the object itself and not to our (perhaps partial or limited) knowledge of it'. In so far as there exist any such properties they are always under some description or other, like the stones in Danto's example. That is to say, their 'intrinsic' (identifying) features are just those features that have been successively picked out over the course of human history, rather than properties which inhere in the object and can thus suffer no alteration under shifts of cultural perspective. This is why Margolis registers such keen disappointment when he finds Kuhn backing away from his earlier (more 'robust') claims concerning ontological relativity and the idea that scientists quite literally inhabit 'different worlds' before and after some crucial paradigm-change. For what he wants from Kuhn – and what he takes Kuhn to have provided despite his later recidivism – is an account of this process which breaks altogether with the notion of science as an enterprise aimed toward producing more accurate and adequate knowledge of an objectively-existing reality.

This notion is simply incoherent, Margolis thinks, since it ends up committed to a physicalist ontology that takes no account of the *history* of science or the various knowledge-constitutive interests that have characterized that history to date. And its limits show up even more clearly when the method is transferred – as by Danto – from the natural-scientific to the social, cultural, and more broadly historical domains. For here its effect is to narrow the range of admissible 'scientific' evidence to just those items of (purportedly) factual knowledge – along with the law-like regularities holding between them – whose 'intrinsic' nature is supposed to provide a guarantee of objectivity and truth. But such a method could bear no resemblance to 'history' in any genuine (humanly meaningful or intelligible) sense of that term. Rather, it would constitute a species of massive and disabling category-mistake, an attempt to legislate for historical discourse in terms of a programme – the deductive-nomothetical or covering-law approach – whose failure is now evident even on its elective home-ground in philosophy of science.[20]

Margolis's point is that we cannot, like Danto, expect to have it both ways, on the one hand adopting a reductively 'physicalist' (objectifying) approach to issues of historical truth while on the other acknowledging – in token fashion – that history must have *something* to do with human motives, purposes, meanings, or intentions. On the contrary: if we want to be 'realist' in the latter respect – to grasp the full extent of human involvement at every stage in the course of scientific or historical enquiry – then we shall have to let go of that false (reified) conception of truth which locates it in the objects or events themselves. *Tertium non datur*, it seems: either a reductive physicalist approach that excludes all concern with meanings, beliefs, attitudes, intentions, or – Margolis's proposal – one that counts objective 'reality' a world well lost for the sake of retrieving those cultural contexts in the absence of which (so he argues) there could be no science or history. Nor does he flinch from accepting, in consequence of this view, a notion that most philosophers at least since Aristotle have regarded as both logically absurd and contrary to all the laws of natural or physical necessity. This is the idea that past events can indeed be altered – subject to a form of reverse (retroactive) causality – since there is nothing in the nature of those events that could render them somehow objectively proof against redescription in radically different (world-transformative) terms.[21]

It may still appear that there is room for a more moderate, less provocative construal of Margolis's claim. On this version he would be arguing merely that interpretations change – along with their associated worldviews – but that they still have reference to a historical past whose events were fixed (or whose truth-values were determined) at the time of occurrence and which therefore cannot *in itself* be altered as a result of subsequent paradigm-shifts. One could then 'charitably' interpret Margolis as not having fallen into a manifest confusion between ontological and epistemological issues, but as simply having chosen to present his argument in dramatic (if somewhat misleading) terms. But Margolis makes it clear that he has no wish for such charity if it means giving up his strong anti-realist stance, that is, his commitment to pressing all the way with an ontological relativism which disallows this convenient escape-route. In fact he views it in much the same light as Kuhn's later retreat in response to those critics who urged him to adopt a more sensible line. Quite simply, according to Margolis, there is no way of upholding the distinction – whether in historiography or philosophy of science – between, on the one hand, those truth-values that pertain to past events or a mind-independent physical realm and, on the other, those varying interpretations that have been or might in future be placed upon it. Which is also to say that a 'realism' with regard to cultural or intentional (humanly salient) properties must therefore entail a corresponding anti-realism with regard to the putative objects and events whose 'world' constitutes the very horizon of past or present enquiry.

Thus 'history [for Margolis] is the Intentionally complex career of real cultural entities, or it is the Intentionally complex characterization of such careers' (*Texts without Referents*, p. 299). From which it follows, as a 'further direct consequence', that, 'granting the reality of cultural phenomena and granting that their real properties would, on a bipolar model of truth-values, yield incompatible attributes or ascriptions of attributes, the ontology here favoured leads us to adopt a relativistic model of truth-like values' (ibid.). Margolis thinks this conclusion unavoidable once the point has been taken – from Kuhn among others – that there *just are* different ('incommensurable') ways of construing the evidence, and hence that the realist argument must collapse for lack of any fixed ontological domain with reference to which one could establish determinate (bipolar) truth-values in any given case. For such values are culturally emergent in the sense of deriving from – and having no existence outside – the history of changing paradigms or worldviews that various observers have brought to them.

Hence, as Margolis sees it, the inevitable failure of Danto's attempt to carve out a realm of objective (historically invariant) truths from the ceaseless flux of events and interpretations. His enterprise founders on the following shoals (I quote Margolis):

(i) the inconsistency of having claimed that the referents of the two [i.e. Roman and Christian) histories are or could be one and the same;

(ii) the impossibility of merely collapsing the ontologies of physical and cultural entities in order to preserve the conceptual distinction of human history itself;

(iii) the impossibility of merely adding extrinsic *relations* involving belief, action, use and the like to 'mere real things' in order to preserve the intended distinction of history;

(iv) the inescapable pertinence of Kuhn's puzzle regarding *the cultural and historical nature of perception itself*, in resolving Danto's question adequately; and, therefore,

(v) the complete failure (on Danto's part) to identify properly *what*, ontologically, are the referents of history (*Texts without Referents*, p. 299).

All of which leaves us (so Margolis thinks) with just one adequate line of response: namely, to abandon this whole misguided endeavour and accept that any 'realism' in the matter of historical or scientific truth-claims will have to be realist primarily in respect of those beliefs, attitudes, or intentions that have made up the history of interpretations to date and which still constitute our sole means of access to the various truths (or 'truth-like values') in question. For we shall otherwise always end up in Danto's unfortunate position. That is to say, we shall confront the impossible task of striving to reconcile a false and reductive objectivity with a needful (though grudging) acknowledgement of the fact that such truths are the product of human attempts to make sense of the world and its forever redescribable objects, events, and properties.

Item (i) above is sufficient, Margolis believes, to dispose of any strong physicalist argument for the existence of realia (objects or facts concerning them) that would justify the ascription of truth-values on a purely extensionalist or belief-independent basis. Item (ii) is somewhat obscurely phrased but would appear to rule out the kind of compromise deal that accommodates 'human history' only in order to preserve some milder ('nonreductive') physicalism whose plausibility rests on its failure to specify what *other* sort of history could possibly exist. Items (iii) and (iv) push this argument further – with assistance from the unreconstructed Kuhn – by rejecting any merely 'relational' account of those various false dualisms (object/subject, nature/culture, content/scheme, intrinsic/extrinsic) that have proved so vexatious in many areas of recent philosophical debate. The relationalist 'solution' is a non-starter, Margolis thinks, since it leaves all the problems firmly entrenched – somewhat after the fashion of Cartesian dualism – by omitting to explain how these realms could 'relate' in any but a vaguely supervenient, parallel, or epiphenomenal mode. Thus item (v) carries the sting of his argument against Danto: that only a robust *relativist* account of the sort that Danto so strenuously resists can allow us 'to identify properly *what*, ontologically, are the referents of history'.

Clearly this sentence entails some pretty drastic redefinition of terms as compared with their normal usage by historians and philosophers alike. 'Ontology' here has nothing to do with such *prima facie* ontological questions as: What objects exist and what are their properties quite apart from our (maybe partial or restricted) knowledge concerning them? What grounds can we have – scientific or other – for ascribing such attributes despite the present lack of observational or epistemic warrant? What can be the justification, historically, for supposing that truth-values exist even in the case of statements which make some determinate claim in respect of some specific past event but whose truth or falsehood cannot be established for want of the relevant evidence? As Margolis deploys it, the term has a more Heideggerian or hermeneutic import, referring to that 'depth-ontological' dimension where objects and events somehow come to light through a process of alethic concealing/revealing that occurs altogether beyond the grasp of mere ontic or conceptual understanding.[22] There is a similar redefinition entailed by Margolis's usage of the term 'referent' in his phrase 'the referent of history'. For on this account, quite simply, there is nothing for historical statements to refer to except those various hermeneutical visions and revisions.

Not that Margolis is by any means in thrall to that mystified 'jargon of authenticity' – that dangerous appeal to potent irrationalist currents of thought – which Adorno diagnosed with pinpoint accuracy in Heidegger's strain of depth-ontological talk.[23] His approach is much closer to the pragmatist reading of Heidegger which has lately gained ground among 'post-analytical' thinkers in quest of some hopeful alternative to logical empiricism and its various successor doctrines.[24] Thus he shares these commentators' general desire to play down the strain of anti-humanist thinking in Heidegger's philosophy, early and late, and to emphasize mainly the treatment of themes such as praxis, intentionality, horizonal awareness, and historically situated being-in-the-world. So it is that Heidegger emerges – improbably enough – as a convert to the broad alliance of interests which Margolis and others perceive to have arisen in the wake of mainstream analytic philosophy.

I have argued elsewhere that this is a version of Heidegger which renders his thinking fit for its own purposes only by filtering out those aspects – among them his express contempt for American pragmatism – that would otherwise constitute a large obstacle to any such proposed alliance.[25] However, it can claim authentic Heideggerian warrant on two counts at least. In Margolis's view they are the 'principal clues' that point toward this current *rapprochement*: namely the well-known statements from *Being and Time* that '[a]n interpretation is never a presuppositionless apprehending of something presented to us', and that '[a]n entity [*Dasein*] for which, as Being-in-the-world, its Being is itself an issue, has, ontologically, a circular structure'.[26] Thus 'ontological' questions are always primarily questions that concern our mode of existence as interpreters whose knowledge of objects, entities, or events takes place against a given cultural horizon – or 'hermeneutic circle' of pre-understanding – in the absence of which such enquiries could make no sense.

This is why Margolis and like-minded commentators see no great problem – differences of technical idiom aside – in reconciling Heidegger's depth-ontological project with the interests and concerns of post-analytical philosophy.[27] In both cases, there is a move toward holistic or contextualist modes of understanding that reject any version of the logical-empiricist attempt to specify the criteria or truth-conditions for particular statements regarding, say, the nature and structure of the physical world or the 'objective' record of historical events. For this programme was based (so it is argued) on a whole set of dubious premises – the 'myth of the given', the idea of transparent epistemic access, the existence of distributed (bivalent) truth-values for any statement with genuine cognitive or veridical content – which have now been shown up as so many products of a false and reductive methodology.[28] Thus Heidegger stands – along with Gadamer, Ricoeur and other exponents of the narrative or hermeneutic turn – as a useful source of alternative wisdom just so long as we discount some of his more extravagant claims with regard to the history of 'Western metaphysics', the epochal withdrawal of Being, and suchlike portentous themes. What remains valid in Heidegger's project is his sensible (quasi-pragmatist) acceptance that *interpretation* goes all the way down, in the natural as in the human or social sciences, and that nothing is lost – and a great deal gained – by renouncing those false objectivist ideals.

AESTHETICS, HISTORY, AND CULTURAL EMERGENCE

I would suggest, on the contrary, that there *are* real losses in this turn toward a hermeneutic (or depth-ontological) paradigm that pushes so far toward erasing any distinction between the various methods or disciplines of truth-seeking enquiry that have developed over a long period in the natural and the human sciences.

One can best see the results of this in Margolis's deployment of arguments from aesthetics and art-history as an instance of the culturally-emergent (hence non-objective) character of all human understanding. Thus he takes the example of Michelangelo's *Pietà* as an artefact that is none the less (in some sense) a 'real entity in our world', has 'real properties' and 'enters into real relationships' (*Texts without Referents*, p. 189). Among its more salient properties, Margolis remarks, is that of 'representing the mourning of Christ's death', an attribute that could doubtless be agreed upon by most (if not all) qualified viewers, past and present, but which can scarcely be thought of as 'objective' or 'intrinsic' to it in the ontological-realist sense. Thus '[i]t is not clear that the property behaves extensionally though it be a "genuine" property; and it is not clear that, if it is a genuine property, there cannot be certain "structural indeterminacies" in the very nature of the sculpture that possesses it' (ibid.). From which Margolis concludes that the 'property' in question is imputed to the artwork through a process of ongoing cultural and intersubjective exchange, a reception-history that alone makes it possible for viewers, critics or art-historians to agree in referring to just that work with just those humanly meaningful attributes. And if this much be conceded, he argues, then we shall have to re-examine a whole range of other deeply entrenched presuppositions concerning the existence of a real-world object domain whose properties we take to be somehow independent of their salience *for us* as culturally situated viewers, observers, or knowers. In particular, we shall have to give up the idea that such properties are uniquely determined by the way things stand in reality, and thus that any statements concerning them will in turn have their content or truth-value fixed by reference to those same things and properties.

Margolis calls this thesis WDT1 and construes it in purely extensionalist terms as follows: 'The world determines truth iff, given the way(s) the world is, for each interpretation of our language there is one and only one way the truth-values can be distributed over our whole sentences as thus interpreted' (p. 188). But the example of the *Pietà* is, he thinks, enough to undermine any such claim not only as applied to aesthetics or art-history (where it could surely find few defenders) but in every field of enquiry where there exists no firm methodological line between issues of truth and issues of cultural context or interpretation-history. Thus: 'on the argument that the physical or natural world is (in one sense, though not in another) an artefact internally posited only within the symbiotized cognitive space of a realist science (rejecting all forms of transparency and privilege), the determination thesis may (arguably) also be adversely affected by an extension of the same concession' (p. 191).

We should not, I think, make too much allowance for the fact that Margolis's claim is hedged around by so many cautious reservations, parentheses and qualifying clauses. For it is sufficiently clear – in this passage and elsewhere – that he sees no room for any 'realist' conception of science and its objects of enquiry except in so far as the latter are contrued as 'artefact[s] internally posited', or as quasi-objects that inhabit a 'symbiotized cognitive space' where the distribution of properties, predicates, or truth-values goes according to their cognitive salience *for us* (as denizens of the cultural lifeworld) rather than their objectively-existing attributes, causal powers, microstructural attributes, and so on. For this is what follows, in Margolis's view, when one 'reject[s] all forms of transparency and privilege', that is to say, all versions of the epistemological paradigm that assumes cognitive access to truths about the way things stand 'in reality'. As with the *Pietà*, so with the 'artefacts' of scientific or historical enquiry: any truth-claims concerning them will always be subject to that process of perpetual, open-ended reinterpretation which comes of their possessing no 'objective' features – no acontextual or culture-independent properties – whereby to reduce their 'structural indeterminacy'. Only if the physicalist programme worked (or perhaps some other, more moderate version of the 'unity-of-science' project) could there be any hope of drawing a line between matters of intersubjective taste or judgement and matters of scientific warrant. But these options are excluded, Margolis believes, by that programme's manifest failure and by the fact – as shown by Kuhn – that scientific observations are always theory-laden and theories themselves always underdetermined by the best evidence to hand.[29]

The reference to Michelangelo crops up again when Margolis tackles Danto on the question of whether those stones can be thought of as possessing any 'reality' aside from the different (i.e. Roman, later Christian, and present-day) cultural contexts in which they have figured under varying descriptions. The passage is worth citing at length since it shows – once again – the extent of his commitment to the strong ontological-relativist case. It also provides a useful point of reference for the various counter-arguments that I shall bring to bear in the rest of this chapter. 'If the stones *are* real, Margolis writes,

> and if the 'different and non-overlapping descriptions' concern the real properties *of* (the real properties *intrinsic to*) the 'stones', then the Roman stones cannot be the same stones as the Christian stones, just as a vase and a bowl made of the same clay are neither the same vase nor the same bowl (the Roman stones being *weights* and the Christian stones *relics*, as Danto tells it). Otherwise, the intentional properties assigned (the ones exhibiting 'aboutness', in virtue of which the Roman and Christian stones – like the *Pietà* and Galileo's pendulum, at least on Kuhn's story – *would have histories*) would not (and could not) be the real properties of the 'mere' stones. They could only be *relationally* associated with those 'mere' stones, by way of *use*, in accord with particular *beliefs* and *actions*. The *stones*, then, could have histories only in a relational, external sense (by association with our beliefs). Histories, then, would only and always be relational and external. But they could not be such for persons and their cultural world, for persons possess beliefs and perform actions: unless, that is, persons, actions, and beliefs can be physicalistically identified and characterized, so that the relational histories required are relativized *to those reduced referents that, by hypothesis, are not intrinsically affected by history.* In short, the failure of physicalism can be shown to entail the paradoxes of history. (*Texts without Referents*, pp. 298–9)

Margolis takes the notion of 'aboutness' from Danto, who uses it mainly in the art-historical context to characterize those salient qualities in certain objects – paintings, sculptures, vases, bowls, and so on – that invite us to treat them as bearing some intentional (purposive or humanly significant) import.[30] The same applies to historical actions and events when considered from Danto's 'internalist' perspective, that is, in so far as they call for the kind of empathetic understanding (*Verstehen*) that enables ascriptions of motive and intent. On his account, this approach has its limits in the context of historical enquiry and certainly cannot be extended to the point where the truth of historical events must be treated on a par – ontologically speaking – with the sorts of validity-claim advanced for judgements of aesthetic status or value. For with respect to historical events it is in the very nature of any such enquiry to presume the existence of a real-world ontic or factical domain whose pastness may render those events more or less opaque, but in the absence of which there could be no distinction between history and the various fictive (or aesthetically appealing) constructions that we might choose to place upon it.

Thus a great deal follows from Danto's argument for the non-reducibility of historical truth to *just* those modes of 'internal', hermeneutic, or intention-based understanding which find their most apt illustration in the case of aesthetic artefacts. What is at issue is the standing of historical discourse as a mode of knowledge that can lay claim to both a genuine object-domain (that is, a field of enquiry that is not just a mirror-image of its own shaping interests and concerns) and to certain well-defined standards concerning the nature, the methods and legitimate scope of such knowledge. Hence the importance, for Danto, of holding the line against a purely internalist approach that would reduce *all* historical enquiry to the 'hermeneutic circle' of interpretative preunderstanding. On this view, quite simply, there is nothing that could count as evidence against some preferred (culturally salient) mode of revisionist emplotment.[31] History becomes just another kind of artefact, subject to all the effects of 'structural indeterminacy' – the absence of objective, mind-independent properties – that characterize a painting like Michelangelo's *Pietà* or other such quasi-'objects' of aesthetic contemplation. To suppose otherwise – so Margolis argues – is to adopt a reductively physicalist approach that takes no account of those intentional (belief-related) features whereby we are enabled to interpret history as a record of humanly-significant actions and events.

Thus Danto's theory requires that we distinguish (say) 'red paintings' and 'mere red canvasses' as belonging to different ontological domains, the one evoking an aesthetic response in virtue of its perceived intentional attributes, the other existing as a physical entity that invites or solicits no such response. But, according to Margolis, this merely evades the issue as to what – 'ontologically' speaking – can possibly serve to make good the distinction on other than internal, intentional, or culturally salient grounds. For if indeed it is the case (*vide* Kuhn) that differing visual perceptions 'belong in different worlds', then there can be no appeal beyond those worlds – or cultural horizons of intelligibility – which could effectively decide the issue from one instance to the next. Danto, he claims, '*never* concedes the point and never addresses it – which, of course, would completely disable his otherwise pretty story of the numerically different but otherwise "perceptually" indistinguishable "red" paintings and "mere red canvasses" ' (*Texts without Referents*, p. 298). This objection applies equally to his (Danto's) bootless endeavour to secure a domain of objective, real-world historical truth against what he sees as the outlook of wholesale revisionist licence entailed by purely 'internalist' approaches. For here also it is Margolis's claim that *nothing* – whether actions, events, paintings, scupltures, vases, bowls, or stones – can be either interpreted, known, perceived, or cognized outside or apart from the various historically emergent cultural contexts wherein they have acquired whatever truth (or meaning) they possess. In short, Danto's argument demonstrably fails for lack of any adequate ontological ground for his realist stance *vis-à-vis* the distinction between 'objective' and other (for example, aesthetic, intention-based, or intersubjectively-validated) orders of truth-claim.

However I think it can be shown that Margolis has himself confused the issues here. With regard to the stones his case goes as follows: that *if* they are 'real', and *if* that reality is the object of 'different and non-overlapping descriptions', then in order for those descriptions to possess any 'objective' (real-world) referential content it must – logically – be a *different content under each of the various descriptions concerned*. But this is just another (superficially plausible) version of the standard anti-realist move which begins by collapsing the ontology/epistemology distinction – that is, by assuming that what is 'real' can only be a matter of what's 'real under this or that description' – and then concludes, by a purely circular process of reasoning, that reality just is (for all that we can possibly know) a product of interpretation. Of course such descriptions may differ – perhaps to the point of radical divergence or Kuhnian 'incommensurability' – in respect of the various features, predicates, properties, or truth-values they assign to some object or event picked out in accordance with some given worldview, conceptual framework, or 'ontological scheme'. But there is no good reason (anti-realist prejudice aside) to conclude from this that different observers are talking about 'different worlds' in the full-blown ontological-relativist sense that *reality itself* changes under varying modes of description. For those observers could either be wrong in certain respects (as with numerous well-documented cases in the history of science), or viewing things from an angle – a cultural perspective – that emphasized certain (to them) salient features and ignored or downplayed certain others. In the scientific context such claims might *purport* to be true – to describe or explain some objective feature of the phyical world – and yet turn out (upon closer investigation) to have misinterpreted the evidence for want of adequate observational techniques or conceptual-explanatory resources. Thus there is no valid argument from the fact that different observers have advanced different truth-claims (or reality-ascriptions) regarding such phenonema to the idea that realism is a self-contradictory thesis since it entails the assignment of diverse (often incompatible) properties or features to a supposedly perduring, self-same physical world. For this is a pure *petitio principii*, taking as read what it sets out to prove: namely that reality *just is* what those various observers have made of it and hence that their contradictory claims entail the non-existence of a real-world domain in respect of which those claims could be held determinately true or false.

One might also question Margolis's idea that aesthetics (or art-history) is the best place to look for useful lessons concerning the intentional, the observer-relative or culturally emergent character of *all* truth-claims in *whatever* discipline or field of enquiry. Thus (to recall): 'the Roman stones cannot be the same stones as the Christian stones, just as a vase and a bowl made of the same clay are neither the same vase nor the same bowl (the Roman stones being *weights* and the Christian stones *relics*, as Danto tells it)' (ibid., p. 298). But to appreciate the aesthetically (and functionally) salient distinction between a vase and a bowl is not to lose sight of the relevant fact – as Margolis lets drop quite contrary to his own thesis – that 'they are made of the same clay', that is, that they possess something in common (a substantive or material property) despite falling under those different descriptions with regard to their formal and cultural attributes. This is yet more evident in the case of the stones, since here it requires a quite extraordinary stretch of the anti-realist (or ontological-relativist) thesis to argue that their change of description – from 'weights' to 'relics' – entails their physical non-identity from one cultural 'world' to the next, or at any rate the lack of any possible criteria by which to establish that identity. All that is needed to avoid this unfortunate outcome is a theory of sortals – of relative identity – which explains how objects can indeed fall under various descriptions (material, formal, functional, aesthetic, sacred, and so on) and yet remain identical for this or that well-defined classificatory purpose[32]. Thus the vase and the bowl might either be described as 'different objects crafted out of the same substance' or 'the same substance crafted into different forms'. Or again, to vary the example, suppose that two different substances had been used in the crafting of a vase (or a bowl) whose formal-aesthetic attributes – mass, shape, proportions, surface texture and so forth – were precisely identical. One would then have the same choice of alternative descriptions but with the terms symmetrically reversed as between 'same/different' and 'substance/form'. What counts is the particular sortal criterion by which such substances or forms are individuated and which hence provides a means of fixing their identity-conditions from one temporal or cultural context to another. Thus different cases may require greatly differing degrees of sortal generality or restrictiveness, as for instance between 'the same as' (applied *sans phrase*) and the specific conditions 'being made of' or 'being crafted into'.

Moreover this allows for tensed applications – or phase-sortals – to cope with the kind of identity-claim that Danto makes about the Roman and Christian stones and which Margolis finds so troublesome. Thus there is reason – indeed compelling reason – to think of them in substance-sortal terms as the same, self-identical stones even though (under a phase-sortal description) they have undergone the change from 'weights' to 'relics'. And one can further specify the difference between phased sortals that apply to natural kinds or instances of organic development and growth – such as acorns growing into oak-trees – and cases like the stones where the criterion of substance-identity is unproblematic (*pace* Margolis), and where the change in question is entirely a matter of cultural context or function. The point about 'relative identity', conceived in these terms, is that it takes full account of the range of divergent descriptions under which some object or event may fall while resisting the currently widespread retreat into forms of holistic, onto-logical-relativist, or extreme anti-realist doctrine.[33]

Where this becomes crucial is in seeking to explain why Margolis (and others) have felt themselves driven to endorse so sweepingly reductive a treatment of scientific and historical truth-claims. The reason is, surely, that Margolis views 'reduction' as *always working the other way around*. That is to say, he views it as always – by very definition – involving the 'physicalist' error which defines truth-values in purely extensionalist terms and which thus precludes any viable account of how knowledge comes about through the constant interaction between subject and object, mind and world, purposive (intentional) human activity and the realm of merely 'ontic' objects and events. But in rejecting this narrowly physicalist approach Margolis swings so far in the opposite direction that he leaves no room for whatever gives *content* to human attitudes, intentions, motives, or beliefs. This is why he makes such a cardinal point of playing Danto's intention-based aesthetic theory against his own (Danto's) more restrictive judgement on the role of intentionalist explanations in matters of historical understanding. For Margolis there is nothing – habit or prejudice aside – that could serve to distinguish these orders of judgement. Since they both involve an 'aboutness' relation (one that cannot be exhausted or adequately described in physicalist terms), therefore both must be considered strictly on a par as regards their ontological status and truth-content. From which it follows that realist ontologies will always, like Danto, run up against the 'paradoxes of history' when they strive to reconcile a falsely objectivist with a properly 'internal' (intention-based) account of objects, events, and their complex careers.

All the same it may be argued that these paradoxes are more of Margolis's than of Danto's creating. That is, they stem from his failure to perceive how the one (physicalist) form of reduction – which Danto is in fact very careful to avoid – cannot be remedied by adopting another, equally reductive approach that rejects realism *tout court* and treats everything – stones included – as products of (or as items internal to) this or that history of cultural redescriptions. Why else should Margolis have such problems with the idea that those stones were undeniably *real* – possessed of certain mind-independent properties like mass, size, hardness, degree of porosity, chemical constitution, molecular and subatomic structure – while also (in virtue of *exactly those properties* or various combinations thereof) playing a humanly significant role in the Roman and Christian, mercantile and religious contexts? Margolis rejects this entire line of argument since he thinks that it involves the realist fallacy of supposing the stones to fall under some one true description which would then be not so much a 'description' – a humanly attributed set of defining properties – as a direct means of epistemic access to whatever constitutes their objective stonehood. But since, as he takes it, any 'fact' about them will itself be a product of human enquiry (scientific, historical, sociocultural, aesthetic, religious-institutional or whatever) therefore we had better just accept that these are all 'intentional' or mind-related properties including those that purport to define what it *is* to be a stone quite aside from such issues of interpretative or cultural context.

This is why Margolis wants to push much further with Danto's idea of the 'aboutness'-relation, the idea that certain objects under certain (e.g. aesthetic or sacred) descriptions may be said to exhibit intentional qualities which elude any physicalist account. Where Danto goes wrong (he argues) is in not seeing that this also applies to every property of every object that has ever been brought under any description whatsoever. Thus Danto makes room for intentional attributes – for beliefs, practices, social contexts, cultural conditions of emergence – but only in so far as they are taken to bear upon a world (past or present) whose reality cannot be reduced to the sum (or the 'intentionally complex career') of those various attributes. To Margolis this looks like another vain attempt to stake out a realm of objectivity and truth by conceding just so much ground and no more to the partisans of cultural relativism. On Danto's account 'they [the intentional attributes] could only be *relationally* associated with those "mere" stones by way of *use*, in accord with particular *beliefs* and *actions*'. And again: 'the *stones* . . . could have histories only in a relational, external sense (by association with our beliefs)' (*Texts without Referents*, pp. 298–9). But this salvage-operation simply won't work, he maintains, if we take the point that those histories 'could not be such for persons and their cultural world, [since] persons possess beliefs and perform actions'. For on his account there is nothing – no mind-independent object-domain – that can be thought of as somehow 'external' to the ongoing process whereby persons both acquire beliefs about the world and act upon it in various ways. Unless, that is, 'persons, actions and beliefs can be physicalistically identified and characterized, so that the relational histories required are relativized *to those reduced referents that, by hypothesis, are not intrinsically affected by history*' (p. 299; italics in original). But anyone who adopted so narrowly physicalist a view of human agents and their modes of interaction with the world would *ipso facto* (he thinks) be at a loss to explain how the course of history – or developments in the natural sciences – could ever themselves be affected in turn by human purposes, motives, or intentions. So it is – to repeat – that 'the failure of physicalism can be shown to entail the paradoxes of history'. For as Margolis sees it there is just no way of saving ontological realism through the adoption of a double-aspect theory (like Danto's) that takes due account of 'internal', hermeneutic, or culturally-emergent features while clinging to an externalist 'myth of the given' as its basis for scientific or historical truth-claims. Thus the only viable (i.e., non-reductionist) means of resolving those paradoxes is one that breaks altogether with such realist – for which read 'physicalist' – dogmas and that treats *every item* in the object-domain, along with all its properties and attributes, as a product of this or that intentional stance taken up with respect to it.

OF POSSIBLE AND ACTUAL WORLDS

This seems to me a very striking example of the difficulties that philosophers have lately got into by fixing their sights on just one chapter of developments in the wake of logical empiricism. Of course there is a sense in which this charge might seem patently absurd when applied to Margolis. After all, his work takes in an enormous range of topics, issues, and thinkers, from the Anglo-American analytic and post-analytic traditions to continental hermeneutics, narrative theory, post-structuralism, deconstruction, and so forth. Moreover, he treats them – as I hope to have treated him – through a detailed close-reading of particular passages in particular texts and not in the breezily name-dropping fashion which Richard Rorty has adopted as a kind of alternative ('post-philosophical') artform.[34] All the same there is another sense in which Margolis appears to have assembled all these diverse tributary sources – Heidegger most prominent among them – in order to escape what he perceives as the dead-end not only of logical empiricism but also of subsequent (for example, Quinean) attempts to change the terms of analytical debate.

Thus Margolis cites the well-known passage from Quine's *Word and Object* where he lays out the programme for a purely extensionalist epistemology that would emulate the (supposed) success of the physical sciences in purging their language of all intensional (likewise inten*tion*al) predicates or expressions. 'If we are limning the true and ultimate structure of reality', Quine writes, 'the canonical scheme for us is the austere scheme that knows no quotation but direct quotation and no propositional attitudes but only the physical constitution and behavior of organisms'.[35] Margolis has no difficulty in showing the paradoxes that arise with any such reductionist account of how enquiry proceeds – or understanding comes about – in the natural sciences, let alone the more hermeneutically oriented disciplines. It is possible, he concedes, that Quine is here registering an 'altogether undefended prejudice'. On the other hand this reading is hard to credit in view of Quine's 'sustained interest in the puzzles of knowledge and science, his persistent efforts over an entire career to entrench his extensionalism, and the seriousness with which his many followers have adhered to his apparent instruction' (*Texts without Referents*, p. 69). Margolis's suggestion is (briefly) that we take the above passage at face value as a statement of the case for a naturalized – science-based or physicalist – epistemology which treats issues of truth, knowledge, and justification as matters of 'first-order' cognitive grasp requiring no appeal to some 'second-order' level of legitimizing theory or principle. Thus, as Quine presents it, this case rests – in Margolis's words – upon 'the plain fact that the requisite first-order competence is taken to be a natural capacity of the members of human communities: to form, codify, discipline, enlarge, confirm, even improve those relatively systematic collections of distributed claims that they are pleased to treat as the core of the various sciences' (p. 69). Margolis is perfectly willing to endorse Quine's position just so long as it involves no further (second-order) epistemological claims concerning the privileged descriptive or explanatory status of the natural sciences. For this would entail all those same paradoxes – of meaning, history, intentionality, agency – that come crowding back as soon as one adopts a wholesale or doctrinaire reductive-physicalist approach.

In short, Quine's programme is acceptable as an account of one way that philosophy has been practised under certain historically-emergent conditions, namely those in which the natural sciences (physics especially) have assumed a high degree of cultural salience. Where it goes off the rails, in Margolis's view, is when the 'seriousness' of Quine's 'many followers' leads them to suppose that the programme could actually be carried through to the point of eliminating all intensional (meaning-related) or intentional (purposive) contexts and predicates. But this can be avoided easily enough by dropping any talk of cognitive or epistemic 'privilege' and substituting talk of cultural 'salience' as sufficient to explain why the programme has enjoyed its recent high prestige among philosophers who like to take the physical sciences as their model of disciplined, constructive enquiry. As Margolis puts it:

> the first-order powers admitted *as* cognitively apt on pragmatist grounds are apt only in the sense of *salience*, not of privilege – that is, only in the sense of what, provisionally, perspectivally, reflexively, in a way internal to the very achievement to be accounted for, appear to be the most promising candidates (for the time being) for the *explananda* required. They are subject to revision and displacement for all sorts of reasons having to do with how salience itself may change.... Put another way, the theme of salience is the theme of the *empirical* grounding of all science and philosophy shorn of any and all first-order or second-order privilege. For example, it permits the acknowledgement of *invariances, universal laws, necessities, essences* within its scope; but, in permitting that, it obliges us to construe their discrimination as encumbered, as ineluctably grounded in whatever is given as salient – as whatever we cannot thus escape. Hence, by a simple maneuver, invariances, essences and the like are *posits* made under and within the tacit, endogenous, incompletely penetrable constraints of salience itself. Anything less would imperil realism (in the pragmatist sense); anything more would restore privilege. (*Texts without Referents*, pp. 70–1)

What is clear from all this – as likewise from his treatment of Danto's philosophy of history – is that Margolis sees no third way (no workable realist alternative) between a thoroughly reductive physicalism in the Quinean mode and a pragmatist approach which views that programme as entailing no ontological commitments beyond those carried by a currently prestigious (science-based) line of talk. For if indeed it is just a matter of cultural 'salience' – of what happens to count, from time to time, as good scientific practice – then of course there can be nothing in the nature of those 'laws', 'invariances', 'necessities' and 'essences' thatwould pertain to the way things stand in reality as distinct from their role within this or that currently favoured descriptive scheme.

Least of all could this offer any support for the kind of causal-realist argument developed by philosophers of science such as Wesley Salmon, Rom Harré and Roy Bhaskar.[36] For on their account it is crucial to maintain the distinction between a realm of real-world (causally-operative) objects, processes, and events and a realm of purposive human activity which may indeed be causally efficacious in its various interactions with the physical world but which none the less calls for a different order of intentionalist (that is, reason-based as opposed to purely causal) explanation. Hence Bhaskar's argument as epitomized in the title of his book *Scientific Realism and Human Emancipation*. That is: by ignoring or sinking the difference between 'intransitive' and 'transitive' domains we risk the twin perils of a thoroughgoing determinism applied to issues of human agency, responsibility, and choice or – conversely – an empty and meaningless notion of human 'freedom' that makes no allowance for those real-world (for example, causal) constraints that in practice set limits to the range of humanly available options.[37] It seems to me that Margolis is open to just this criticism when he suggests that we take Quine's programme at its word but then reconstrue it in terms of cultural 'salience', or as happening to fit with the elective self-image of a given interpretative community. For on this view there is nothing that cannot be thus redescribed, from the quasi-universal 'laws', 'invariances', or 'necessities' of (a certain) scientific language-game to the course of past events as viewed from (a certain) cultural-historical perspective. More precisely: any constraints upon this process will have to do with their relative salience *for us* as culturally situated knowers or observers, rather than with their truth-content as a matter of objective (scientific or historical) fact.

We have seen already just how far Margolis is prepared to push this argument when it comes to the issue – the 'peculiarly treacherous' issue – of revisionist historiography. While acknowledging that these are deep waters ethically and politically he none the less advances some strong (in the literal sense *preposterous*) theses with regard to the 'flux of history', or – what he takes to follow from this – the malleability of past actions and events in consequence of changes in our present understanding of them. Here again Margolis makes a point of rejecting any moderate construal of his argument that would view it in merely 'relational' terms, that is, as holding (1) 'that it [the historical past] can acquire new *relationships* with an evolving present', and (2) that 'we can always abandon, for cause, a previous *characterization* wrongly taken as historically true, on the basis of new, even hitherto nonexistent, evidence' (*Texts without Referents*, p. 293). For this would amount to just a Danto-style compatibilist account that left some room for shifts of cultural horizon or intentional perspective while offering no challenge to the standard view according to which, 'if it is real, then the past is fixed and unchangeable'.

However, there are compelling reasons – logical and metaphysical as well as ethico-political – for holding that this is indeed the case and that any idea of retroactive causation will give rise to insoluble aporias. These reasons were first set forth by Aristotle in his discussion of the sea-battle which *either would or would not* take place at some specified future date, and whose occurrence or non-occurrence (along with its outcome and everything else about it) might thus be thought of – mistakenly – as determined in advance. In Aristotle's view this mistake comes about through a failure to distinguish the two orders of logical and causal-empirical necessity, which in turn gives rise to a kindred confusion with regard to past events and future possibilities. Thus the one domain is 'closed' in the sense – quite simply – that it consists of events that have *already occurred* irrespective of what we don't yet know or might yet discover concerning them. The other is 'open' in the sense that we can know nothing about it save what is more or less reliably predictable by induction from past experience plus the kind of logically valid but empirically vacuous either/or disjunction that applies in the case of the sea-battle.[38]

Later philosophers developed and refined this argument in various ways, whether (like Leibniz) through metaphysical enquiries into modal logic and the orders of relationship between actual and possible worlds or (like Kant) through epistemological reflection on the scope and limits of human knowledge. At present the debate is most active in the field of particle physics where the idea of simultaneous (i.e., faster-than-light) quantum action-at-a-distance would appear to be borne out both in theory and by certain deeply puzzling experimental results.[39] However these results are still a topic of widespread dispute and in any case cannot be taken to support extrapolations from the micro- to the macrophysical domain. With regard to the latter it is still necessarily the case that past events are characterized by their having already occurred, and are thus subject to no possible form of retroactive causal influence, while future events are either contingent upon certain establishe dregularities holding good or else open to various outcomes depending on a vast (unpredictable) range of causal and other (for example, purposive or human agency-related) factors. As Richard Swinburne puts it: 'the concepts of past and present cannot be connected to the rest of our conceptual scheme unless we understand the past as the logically contingent that is causally unaffectible, and the future as the logically contingent that is causally affectible'.[40]

Swinburne goes on to examine various arguments for the possibility of backward causation, among them an ingenious thought-experiment proposed by Michael Dummett.[41] Where they all break down – Dummett's included – is at the point when one asks what could possibly count as *evidence* for past effects having come about through the action of present causes. For it then turns out that the terms 'past' and 'present' must either be applied in their normal senses (in which case the claim becomes simply incoherent) or else have been redefined somewhere along the way so as to switch meanings (in which case the claim self-evidently fails for lack of argumentative content). Thus Dummett's imaginary world is merely one in which temporal predicates are reversed and history (so to speak) runs backwards without otherwise affecting anything in the nature of logical or causal necessity. And the same goes for any modified or strengthened version of Dummett's experiment which attempts to get around this problem by supposing a mirror-image of our own world (that is, the world of forward-acting causation) in which 'all causes occur later than their effects and are linked to them by causal chains'. In the world thus envisaged '[b]efore an event happens you cannot prevent it from happening, but you have much information at most times about what will happen. After it happens, however, you can affect whether or not it happens'.[42]

This world has a certain logical consistency and at any rate avoids the more obvious problems with backward causation as a claim applied to an otherwise forward-moving temporal frame of actions and events. However, it secures this tactical advantage at the cost (once again) of redefining terms in a manner which effectively deprives the thesis of all substantive content. 'It is still incoherent', Swinburne writes,

> if you take the 'later' and 'earlier' in the description of it as having the same meaning as in ours; for the meaning of these terms, as we use them, is such as to allow the logical possibility of a cause having a later effect. But if you let such terms acquire their meaning from the description of the imaginary world alone, then

that world turns out to be the same as ours – its 'later' is to be read as our 'earlier', etc.[43]

Margolis will hardly need reminding of these logical and metaphysical arguments against the possibility of backward causation. Indeed he is careful, at times, to let in certain parenthetical or qualifying clauses which suggest that his proposal is not – after all – so wildly counter-intuitive since it has to do with our changing *interpretations* of history rather than with historical events as caused (or influenced) by our present understanding of them. All the same these concessions are more often countermanded by a strong-revisionist *ontological* mode of argument which, as we have seen, goes out of its way to block off any such tactical retreat. Thus, '[t]he ontic indeterminacies of history are simply the ontic indeterminacies of culturally emergent entities and phenomena – indeterminacies due to their intrinsically possessing intentional properties' (*Texts without Referents*, p. 302). And again, more bluntly: 'the "historical past" can be "affected" in a realist sense by present historical novelty'. (Any hint of retreat in those uneasy quotation-marks is promptly withdrawn by his emphatic phrase 'in a realist sense'.) From which it must be taken to follow, *contra* Swinburne, that 'the *time* of history is . . . always the present of human cognition and interest (or the past posited from that present') (p. 305).

These are 'treacherous issues', as Margolis readily acknowledges, because they bear not only on the question of historical truth and its adaptability to present (e.g., ideological) purposes but also on the related ethical question of how far human beings can be held responsible for their own actions, past and present. For this latter concerns the extent to which actions are thought of either as always already causally determined or as remaining open – in some degree at least – with respect to an as-yet unforeclosed range of choices, decisions, or commitments. What is crucial here, as Swinburne notes, is the joint operation of causal and intentional factors in bringing about certain changes in the world (and ourselves) which cannot be explained on any purely physicalist or reductionist account. He is thus very much in agreement with Margolis on this point at least: that physicalism – along with its kindred (extensionalist) programme in philosophical semantics – is a doctrine that signally fails to capture what is most salient about human actions, motives, meanings, purposes, and intents. But he comes out very strongly against any argument – like Margolis's – which takes that failure as a sure sign not only that physicalist approaches are altogether bankrupt in this regard but also that causal-explanatory factors can have no place in the realm of history or other such 'intentional' (action-related or humanly meaningful) domains. For we should lack all sense of the operative difference between *acting* and *being acted upon* were it not for our existing as real-world agents subject to various kinds and degrees of causal constraint, among them – crucially – the impossibility of altering past events in accordance with our present wishes and desires.

Moreover, there is a close (indeed inextricable) relation between intentional activities such as *trying*, *attempting* or *meaning-to*, and the deployment of those human causal powers that affect the outcome – successful or otherwise – in any given case. For '[i]t is in virtue of exercising causal powers ourselves, and being aware of ourselves as doing so, that we acquire the concept of causation'.[44] And having acquired that concept we then have a basis for distinguishing (say) the non-intentional effects upon us of various physical causes; the extent of such causal factors as they bear upon our own and other people's actions; the range of intended and unintended consequences that may follow from those actions; the degree of moral accountability involved in cases where behaviour (and intention) are thereby subject to certain kinds of causal constraint; and the power that we and others none the less exercise to alter or expand the possibilities of future action by adopting some particular (causally effective) choice among the various alternatives presently on offer. However we could have no grasp of these crucial distinctions were it not for our sense of the limits placed upon human freedom by the existence of certain real-world constraints – such as physical causality and (*pace* Margolis) the 'fixed, unchangeable' character of past events – in whose absence 'freedom' would be a wholly meaningless idea. In other words there is nothing contradictory about adopting a causal-realist position with regard to both physical objects or events and the role of human agency while also maintaining that actions may be free – causally underdetermined – in respect of their motives, purposes, or intended outcome. Indeed, so far are these arguments from getting into conflict that they provide the only adequate joint account of how human beings are able to exert some significant (willed and causally efficacious) influence on the course of events.

They are also, as I have said, very much involved in our judgements of moral responsibility and the allowance we make for those various causal factors that operate either to enhance or to restrict the human capacity for autonomous choice and action. In Swinburne's words:

> There is a distinction, of which agents are aware in themselves, between intentionally bringing about some effect and intentionally permitting it to occur; between intervening in nature and allowing nature to take its course; between exercising active causal influence and refraining from doing so. Given that the agent is not under any natural or causal necessity to refrain, his refraining constitutes him as a permissive cause of the effect which he allows to occur.[45]

It scarcely needs pointing out how acute is the relevance of these and kindred distinctions in the legal, ethical, and (especially) the medical context where they may – for example – make all the difference between judgements or verdicts of 'killing' and 'letting die'. But the general point is more basic than this and has to do with the very possibility of construing human actions and purposes in humanly intelligible terms. What Margolis at times seems to envisage is a degree of 'intentional' control over acts and events, past and present, which surpasses any possible human capacity and approximates more to the attributes of divinity as defined in Christian theological doctrine. Here again some of Swinburne's remarks are very much to the point. For his arguments about time, causation, agency, and intentionality are set out by way of metaphysical prolegomenon to a theodicy whose main tenets derive from a series of contrastive definitions concerning God's omniscience, omnipotence, temporal transcendence, and so forth. That is to say, he puts the case that any adequate defence of Christian doctrine will also require an adequate account of those distinctively human attributes whose scope (and whose limits) are in part determined by our mode of existence as physically embodied creatures exercising causal and intentional powers under certain specifiable real-world constraints.

It seems to me – though he would no doubt dissent very strongly – that we can have Swinburne's metaphysics and ontology without his theodicy. In particular we can take his point about the manifold confusions that result from ignoring those real-world (causal and temporal) constraints and thus entertaining extravagant ideas about the malleability of past events or the power that we exercise, as intentional agents, to determine what shall count as a 'law of nature' or a matter of established fact. The following passage lays his argument out most succinctly:

> I conclude that the existence of a substance who has necessarily pure, intentional, limitless power entails and is entailed by the existence of a substance who has necessarily the divine properties [here] described. . . . To have that is his nature. The claim that there is a God is therefore to be read as the claim that there is such an individual. It follows not merely that no individual who is not divine could become divine, but also that no individual who is not divine could ever (even everlastingly) have been divine. For being divine is essential to the individuals that are divine, and is part of what makes them the individuals they are.[46]

I have no wish to pursue Swinburne any further into the intricate details of his argument concerning the Christian doctrines of incarnation, atonement, the virgin birth, and so forth. Sufficient to say that they all turn on the distinction between divine attributes and the various temporal, physical, and causal factors that constitute the very conditions of possibility for human knowledge and experience. Most importantly they help to show what is wrong with those less guarded statements of Margolis that can only be construed as endorsing a full-scale (ontological) version of the case that reality *just is* a cultural construct out of our various language-games, narrative paradigms, or intentional visions and revisions.

From Swinburne's point of view such claims would bear witness to a massive confusion of onto-metaphysical realms, a divinization of human powers and capacities that produces all manner of absurdity, paradox, and (not least) theological error. However, for present purposes his argument can be stated without these severe doctrinal sanctions attached. Again it is focused most clearly with regard to the issues of causality, intention, and temporal sequence since it is here (according to Swinburne) that human understanding is ineluctably subject to conditions that preclude the ascription of divine or quasi-transcendental attributes.

> If someone says that there are causal processes operating simultaneously in a contrary temporal direction to other processes, the question is – can they (logically) interact with each other or not? If they can, there is the possibility of an event preventing the occurrence of its cause and so its own occurrence; if they cannot

there seems no content in the claim that the processes are simultaneous. I conclude that all causal processes must operate in the same temporal direction, and since some causes are earlier than their effect, all must be. The logical impossibility of backward causation is explained by the causal theory of time – the later just is the causally affectible.[47]

Swinburne also has a good deal to say about the causal (Kripke/early Putnam) theory of reference, naming and necessity. This theory he regards – justifiably I think – as lending weight to his argument for the status of certain unalterable truths about the way things are (or have been) in respect of a physically existent world and the way that we have acquired knowledge of them through causal-epistemic interaction with that world.[48] Thus: 'what makes something water is not its taste, its transparency, its density, etc., but its chemical composition (H_2O) which underlies and is causally responsible for the observable properties'.[49] He is by no means so confident that any such underlying essence (or clear-cut identity criterion) can be offered in the case of human beings, whether in virtue of their genetic constitution, their physiology, anatomy, capacities for reason, for purposive (intentional) action, or whatever. In fact Swinburne runs through most of the criteria proposed by thinkers from Aristotle down and finds them either lacking in some crucial regard as measured against our normal conception of personhood or subject to counter-factual hypotheses that throw their adequacy into doubt. As might be expected his argument tends toward a concept of the human as intrinsically defined by the possession of a soul whose necessary ('logical') attributes include its continuing identity over time and its non-susceptibility to accidents of physical change or deprivation. But again this leaves a fair amount of room for discounting Swinburne's more overt doctrinal or theologically-sanctioned commitments. Thus: 'how much by way of powers an individual has to gain or lose in order to cease to be human is unclear. But it follows from my earlier arguments that I (and any other human) who gain or lose powers or body am essentially a soul, someone with a capacity for feeling, thought, or intentional action'.[50] And it is precisely these latter attributes – 'soul' apart – that when taken along with his causal theory of actions and events constitute Swinburne's persuasive case for a realist metaphysics of knowledge and experience.

CONSEQUENCES OF ANTI-REALISM

I have focused on Margolis's treatment of these issues not because he is a soft target but because, on the contrary, he is among the most subtle, resourceful and articulate proponents of what is nowadays a widely influential movement in philosophy of science and inter-pretation-theory. Its central premise is the need to abandon those 'physicalist' or truth-theoretic (for which read: logical-empiricist) doctrines which, until recently, occupied the high ground of Anglo-American debate. Hence, for instance, Nelson Goodman's in-genious restatement of the Humean puzzle of induction, based on the use of non-natural predicates ('grue', 'bleen' etc.) which are taken to block any possible appeal to law-governed regularities or causal explanations.[51] However, it is wrong – artificially restrictive – to suppose that some variant of logical empiricism is the only option for anyone in quest of a realist theory of truth, knowledge, and representation or an account of science that would offer an alternative to the cultural-relativist line.

'Needless to say', Margolis writes, 'the first-order powers ad-mitted as cognitively apt on pragmatist grounds are apt only in the sense of *salience*, not of privilege – that is, only in the sense of what, provisionally, perspectivally, reflexively, in a way internal to the very achievement to be accounted for, appear to be the most promising candidates (for the time being) for the *explananda* re-quired' (*Texts without Referents*, p. 70). To reject this argument – along with its cultural-relativist upshot – is to fall straight back into naive 'physicalism' on the one hand (i.e., the notion that knowledge has to do with a realm of intransitive, mind-independent objects, processes, or events) and on the other an equally naive idea that the logic of enquiry entails certain standards or validity-conditions that somehow transcend our localized habits of belief or evaluative judgement. But surely it is the case, as Margolis remarks in Quinean–Kuhnian vein, that those standards are themselves 'sub-ject to revision or replacement for all sorts of reasons having to do with how salience itself may change' (ibid.). From which he derives the now-familiar lesson: that 'the importance of admitting salience is just that it precludes privilege and acknowledges the profound transience and contingent stability of what, in terms of the implicit consensus of actual societies pursuing enquiries of high discipline, *appears* to them to be their science and cognitive power' (p. 70). Thus there can be no question of science providing a special (epistemically 'privileged') example of the way that advances in knowledge come about through procedures of ever more elaborate theory-construction and more adequate empirical testing. Any appearance of progress in either regard must be viewed as just a sign of what counts for us – by our own cultural or evaluative lights – as good in the way of belief. Least of all can such knowledge have anything to do with the real-world, objective properties or attributes of those various phenomena that science sets out to investigate. For on Margolis's account it is only in virtue of their acquiring cultural 'salience' – of their coming to count as worthwhile topics of enquiry for this or that community of knowers – that such properties emerge from the otherwise inchoate and ceaseless flux of experience.

It seems to me that Margolis is led to adopt this position by his fixed antipathy toward realist philosophies of science and by his failure to grant that not all such philosophies come down to some form of reductive physicalist or hardline determinist creed. We have seen already – in connection with Danto and the issue of historical understanding – how far Margolis is willing to go with a strong interpretivist thesis which renders past events (along with material objects like the 'Roman' and 'Christian' stones) subject to a process of open-ended cultural redescription, thus leaving no room for such otiose ideas as historical truth or perduring physical reality. At various points in his book one can detect the slide from a moderate version of this thesis – one which preserves an (albeit weakened) ontology-epistemology distinction – to a wholesale version that places no limits on the scope for redescribing, and hence transforming, the putative 'objects' of scientific or historical enquiry. Thus there are, he writes, 'very good reasons for supposing that the world of human culture at least – chiefly, persons, artifacts, artworks, words and sentences, actions, institutions, and the like – *are* ontologically affected, as the real phenomena they are, in the manner [here] sketched' (*Texts without Referents*, p. 189). One might remark of this passage that the argument is carried on a series of less-than-perspicuous distinctions between, for example, the 'real' and the 'phenomenal', or 'ontology' construed both in realist terms and with suggestions of a different, Heideggerian or depth-hermeneutical import. It is this ambivalence that saves the sentence from flat contradiction and which opens the way to Margolis's next, more extreme formulation of the anti-realist claim. Thus: 'if this much be conceded, then, on the argument that the physical or natural world is itself (in one sense, though not in another) an artifact internally posited only within the symbiotized cognitive space of a realist science (rejecting all forms of transparency and privilege), the determination thesis may (arguably) also be adversely affected by an extension of the same concession' (ibid.). Again there is a strong sense that Margolis is hedging his bets, taking refuge – like the later Putnam – in a quasi-Kantian notion of 'internal realism' that would somehow leave *ultimate* 'reality' untouched while denying any epistemic (humanly available) access to such reality from within the 'symbiotized cognitive space' where 'realist' science is just another artifact of salient cultural interests.[52] However, the point can be made more simply with regard to his treatment of aesthetic and art-historical issues. What the case here amounts to – passing concessions aside – is a principled denial that scientific or historical truth-claims are in any way 'essentially' distinct from the kinds of revisionist interpretative approach that apply to the culturally emergent meanings of an artwork like Michelangelo's *Pietà*. To think otherwise is merely to display one's attachment to an outworn 'physicalist' paradigm that lòcates the truth-value of statements about stones, pendulums, physical artefacts, or past events in a realm that could somehow be held distinct from our culturally 'salient' scientific interests or habits of historical belief-evaluation.

When Margolis talks about 'transparency and [epistemic] privilege' as the two chief dogmas of physicalism one may suspect that his real, if unacknowledged, target is not so much the truth of particular claims in various disciplines of knowledge but a residual, vaguely Cartesian notion of such claims as grounded in the direct access to clear and distinct ideas. The result – here as so often in recent debate – is to polarize the issue so that *all* claims-to-truth come out as pretty much on par, that is to say, as adopting (in Rorty's phrase) a delusory 'God's-eye view' that ignores the evidence of historical change, varying patterns of cultural 'salience', scientific paradigm-shifts, and so forth. However, one may readily accept all this as a matter of well-documented fact while none the less remarking that those various disciplines have developed standards of enquiry – validity-conditions, critical procedures, methods of verification or falsification – which alone make it possible to explain how such changes came about not merely (as Rorty would have it) through a process of random linguistic and cultural drift but through the effort to achieve a more adequate

understanding of certain (perhaps previously obscure or unex-
plained) phenomena. Of course the validity-conditions differ
widely as between, say, research in the physical sciences, in epistemology, philosophical semantics, or the realm of historical
enquiry. Most crucial here is the relative weighting of causal explanations in science – taken to exclude (or at any rate to marginalize)
any 'transitive', mind-dependent, or humanly meaningful dimension – and in historical research where, as Danto allows, these latter
sorts of interest will usually play some role in the attempted reconstruction of significant episodes and events. However, such enquiry could make no progress – would lack any distinctive or
well-defined object domain – were it not for the existence and perduring relevance of certain material facts about history (as witnessed, for instance, by those famous stones) whose reality
Margolis would all but deny through his espousal of the strong-
revisionist or hermeneutic viewpoint.

Thus Danto is fully justified – and in no sense prey to some
reductive physicalist prejudice – when he claims that 'reference
makes realists of us all', and moreover that '*historical* reference is
simply reference in a certain temporal direction relative to the
referring expression itself'.[53] Of course this is a pretty bare-bones or
minimalist description of what historical discourse is all about. But
it gives no warrant for Margolis's charge that Danto is unconcerned
to 'sort out ontologically appropriate objects', or that his theory
entails 'a seamless conceptual connection between physical science
and human history' (*Texts without Referents*, p. 295). For it is just
Danto's point – as against other, more extreme versions of the
hermeneutical approach – that there is a heavy price to pay for
abandoning these crucial distinctions. That is to say, if *everything* is
viewed from an interpretivist perspective (objects, events, causes,
actions, reasons, motives, and so forth) then indeed there can be
nothing to choose ontologically between science and pseudo-
science, or historical truth and whatever can be passed off as such
according to current revisionist needs. This is scarcely an outcome
that Margolis would welcome, given his trenchant criticisms of
Goodman and – more urgently – his express reservations with
regard to its ethico-political consequences. Still it is hard to see
what effective counter-arguments he could muster unless by
adopting a very different view of the truth about relativism, the
various more cogent alternatives to it, and (not least) the sources of
its presently widespread cultural appeal.

Notes

1. Minimalist Semantics and the Hermeneutic Turn

1. See for instance Richard Rorty, 'Overcoming the Tradition: Heidegger and Dewey', in *Consequences of Pragmatism* (Brighton: Harvester, 1982), pp. 37–59; also Rorty, *Essays on Heidegger and Others* (Cambridge: Cambridge University Press, 1991).
2. Mark Okrent, *Heidegger's Pragmatism: Understanding, being, and the critique of metaphysics* (Ithaca, NY: Cornell University Press, 1988).
3. Hubert L. Dreyfus, *Being-in-the-World: A commentary on Heidegger's Being and Time, Division I* (Cambridge, Mass.: MIT Press, 1991).
4. See especially R. Carnap, 'The Elimination of Metaphysics Through Logical Analysis of Language', in A.J. Ayer (ed.), *Logical Positivism* (New York: Free Press, 1959), pp. 60–81; also Carnap, *The Logical Structure of the World* (London: Routledge & Kegan Paul, 1937) and *Meaning and Necessity* (Chicago: University of Chicago Press, 1956).
5. W.V. Quine, 'Two Dogmas of Empiricism', in *From a Logical Point of View* (Cambridge, Mass.: Harvard University Press, 1953), pp. 20–46; also Quine, 'Speaking of Objects', in *Ontological Relativity and Other Essays* (New York: Columbia University Press, 1969), pp. 1–25; 'Posits and Reality', in *The Ways of Paradox and Other Essays* (New York: Random House, 1966), pp. 233–41; *Word and Object* (Cambridge, Mass.: MIT Press, 1960).
6. See Rorty, *Consequences of Pragmatism* (op. cit.) and *Philosophy and the Mirror of Nature* (Oxford: Blackwell, 1980); also Rorty, 'Is Truth a Goal of Enquiry? Davidson *versus* Wright', *Philosophical Quarterly*, Vol. 45 (1995), pp. 281–300; Donald Davidson, 'The Structure and Content of Truth', *Journal of Philosophy*, Vol. 87 (1990), pp. 297–328; Crispin Wright, *Realism, Meaning and Truth* (Blackwell, 1987); Paul Horwich, *Truth* (Blackwell, 1990); Lawrence E. Johnson, *Focusing on Truth* (London: Routledge, 1992); Richard L. Kirkham, *Theories of Truth: A critical introduction* (Cambridge, Mass.: MIT Press, 1992).
7. Alfred Tarski, 'The Concept of Truth in Formalized Languages', in *Logic, Semantics and Metamathematics*, trans. J.H. Woodger (London: Oxford University Press, 1956), pp. 152–278.
8. J.E. Malpas, *Donald Davidson and the Mirror of Meaning* (Cambridge: Cambridge University Press, 1992), pp. 246–7.
9. Ibid., pp. 247–8.
10. See the essays collected in Donald Davidson, *Inquiries into Truth and Interpretation* (Oxford: Clarendon Press, 1984).
11. See references to Quine, note 5, above; also Thomas S. Kuhn, *The Structure of Scientific Revolutions* (2nd edn, Chicago: University of Chicago Press, 1970).
12. Davidson, 'On the Very Idea of a Conceptual Scheme', in *Inquiries into Truth and Interpretation* (op. cit.), pp. 183–98.

13.   See Quine, *Word and Object* (op. cit.); also 'Indeterminacy of Transla-
       tion Again', *Journal of Philosophy*, Vol. 84 (1987), pp. 5–10; Donald
       Davidson, 'Radical Interpretation', in *Inquiries into Truth and Interpre-
       tation* (op. cit.), pp. 125–40; F. Günthner and M. Günthner-Reutter
       (eds), *Meaning and Translation* (London: Duckworth, 1978); Robert
       Kirk, *Translation Determined* (Oxford: Clarendon Press, 1986); David
       Lewis, 'Radical Interpretation', in *Philosophical Papers*, Vol. 1 (Lon-
       don: Oxford University Press, 1983); Colin McGinn, 'Radical Inter-
       pretation and Epistemology', in Ernest Lepore (ed.), *Truth and
       Interpretation: Perspectives on the philosophy of Donald Davidson*
       (Oxford: Blackwell, 1986), pp. 356–68; Talbot J. Taylor, *Mutual Mis-
       understanding: Scepticism and the theorizing of language and interpreta-
       tion* (London: Routledge, 1992).

14.   Benjamin Lee Whorf, *Language, Thought and Reality* (Cambridge,
       Mass.: MIT Press, 1956).

15.   Kuhn, *The Structure of Scientific Revolutions* (op. cit.); see also Quine,
       Theories and Things (Cambridge, Mass.: Harvard University Press,
       1981); Harold Brown, *Perception, Theory, and Commitment: The new
       philosophy of science* (Chicago: University of Chicago Press, 1977);
       Richard E. Grandy (ed.), *Theories and Observation in Science* (Engle-
       wood Cliffs, NJ: Prentice-Hall, 1973); Sandra G. Harding (ed.), 'Can
       Theories be Refuted?', *Essays on the Duhem-Quine Thesis* (Dordrecht:
       D. Reidel, 1976); David Papineau, *Theory and Meaning* (London:
       Oxford University Press, 1979); Michael E. Levin, 'On Theory-
       Change and Meaning-Change', *Philosophy of Science*, Vol. 46 (1979),
       pp. 407–24; Ernest Nagel, Sylvain Bromberger and Adolf Grün-
       baum,*Observation and Theory in Science* (Baltimore: Johns Hopkins
       University Press, 1971).

16.   See for instance Quine, *Pursuit of Truth* (Cambridge, Mass.: Harvard
       University Press, 1990); Christopher Hookway, *Quine: Language,
       experience, and reality* (Cambridge: Polity Press, 1987); Robert Barrett
       and Roger Gibson (eds), *Perspectives on Quine* (Oxford: Blackwell,
       1989).

17.   Davidson, 'The Structure and Content of Truth' (op. cit.).

18.   See note 7, above.

19.   Malpas, *Donald Davidson and the Mirror of Meaning* (op. cit.),
       p. 271.

20.   Davidson, *Inquiries into Truth and Interpretation* (op. cit.), p. 198.

21.   Davidson, 'A Nice Derangement of Epitaphs', in R. Grandy and
       R. Warner (eds), *Philosophical Grounds of Rationality: Intentions,
       categories, ends* (London: Oxford University Press, 1986), pp. 157–74.

22.   Davidson, *Inquiries* (op. cit.), p. 198.

23.   Quine, 'Two Dogmas of Empiricism' (op. cit.), p. 41.

24.   Davidson, 'A Nice Derangement of Epitaphs' (op. cit.), p. 173.

25.   For this reading of Derrida, see S. Pradhan, 'Minimalist Semantics:
       Davidson and Derrida on meaning, use, and convention', *Diacritics*,
       Vol. 16 (Spring 1986), pp. 66–77; also Samuel S. Wheeler, 'Indetermi-
       nacy of French Translation: Derrida and Davidson', in Lepore (ed.),
       Truth and Interpretation (op. cit.), pp. 477–94.

26. Jacques Derrida, 'Signature Event Context', *Glyph*, Vol. 1 (Baltimore: Johns Hopkins University Press, 1977), pp. 172–97; John R. Searle, 'Reiterating the Differences', ibid., pp. 198–208.

27. Christopher Norris, *Derrida* (London: Fontana, 1987); see also my discussion of these matters in Norris, 'Reading Donald Davidson: Truth, meaning, and right interpretation', *Deconstruction and the Interests of Theory* (London: Pinter Publishers, 1988), pp. 59–83.

28. Rorty, *Philosophy and the Mirror of Nature* (op. cit.).

29. See especially Rorty, 'Introduction: Pragmatism and Philosophy', in *Consequences of Pragmatism* (op. cit.), pp. xiii–xlvii.

30. Norris, 'Reading Donald Davidson' (op. cit.). See also Rorty, 'Is Truth a Goal of Enquiry: Davidson *versus* Wright' (op. cit.); Maria Baghramian, 'Rorty, Davidson and Truth', *Ratio*, Vol. 3 (1990), pp. 101–16.

31. See for instance Martin Heidegger, *Being and Time*, trans. John Macquarrie and Edward Robinson (Oxford: Blackwell, 1962); *Basic Writings*, ed. D.F. Krell (London: Routledge, 1977); *Early Greek Thinking*, trans. D.F. Krell and Frank Capuzzi (New York: Harper & Row, 1975); *An Introduction to Metaphysics*, trans. Ralph Mannheim (New Haven, Conn.: Yale University Press, 1959); *On the Way to Language*, trans. Peter D. Hertz (Harper & Row, 1971); *What Is Called Thinking?*, trans. Fred D. Wieck and J. Glenn Gray (Harper & Row, 1968).

32. Friedrich Nietzsche, 'Of Truth and Falsity in their Ultra-Moral Sense', *Complete Works*, Vol II., ed. Oscar Levy (London: Allen & Unwin, 1911).

33. Thus on the one hand Davidson agrees with Rorty that 'nothing counts as justification unless by reference to what we already accept', with the consequence that – in this ultimate sense – 'there is no way to get outside our beliefs and our language so as to find some test other than coherence'. But on the other hand, he takes it as self-evident – *contra* Rorty – that 'we nevertheless can have knowledge of, and talk about, an objective public world which is not of our making' (Rorty, 'Pragmatism, Davidson and Truth' and Davidson, 'A Coherence Theory of Truth and Knowledge', in Lepore [ed.], *Inquiries into Truth and Interpretation*, op. cit., pp. 333–5 and 307–19). See also Davidson's lectures on 'The Structure and Content of Truth' (note 6, above) and Frederick Stoutland, 'Realism and Anti-Realism in Davidson's Philosophy of Language', Parts 1 & 2, *Critica*, Vol. 14 (1982), pp. 13–51 and 19–47.

34. For a useful discussion of these issues see Simon Evnine, *Donald Davidson* (Cambridge: Polity Press, 1991).

35. Quine, 'Two Dogmas of Empiricism' (op. cit.); Nelson Goodman, *Fact, Fiction and Forecast* (Cambridge, Mass.: Harvard University Press, 1955) and 'The Infirmities of Confirmation Theory', *Philosophy and Phenomenological Research*, Vol. 8 (1947), pp. 149– 51); Adolf Grünbaum and Wesley C. Salmon (eds), *The Limitations of Deductivism* (Berkeley & Los Angeles: University of California Press, 1988); Donald Gillies, *Philosophy of Science in the Twentieth Century: Four central themes* (Oxford: Blackwell, 1993).

36. Malpas, *Donald Davidson and the Mirror of Meaning* (op. cit.), p. 269.

37. Heidegger, *What Is a Thing?* (op. cit.).

38. W.H. Newton-Smith, *The Rationality of Science* (London: Routledge & Kegan Paul 1981), p. 173.

39. See especially James Robert Brown, *The Laboratory of the Mind: Thought experiments in the natural sciences* (London: Routledge, 1991) and *Smoke and Mirrors: How science reflects reality* (Routledge, 1994); also Roy Sorensn, *Thought Experiments* (New York & London: Oxford University Press, 1992); Peter L. Gallison, *How Experiments End* (Chicago: University of Chicago Press, 1987).

40. See Robert Ackerman, *Data, Instruments, and Theory: A dialectical approach to the philosophy of science* (Princeton, NJ: Princeton University Press, 1985); Ian Hacking, *Representing and Intervening* (Cambridge: Cambridge University Press, 1983); Don Ihde, *Instrumental Realism: The interface between philosophy of science and philosophy of technology* (Bloomington, Ind.: Indiana University Press, 1991).

41. Karl R. Popper, *Quantum Theory and the Schism in Physics* (London: Hutchinson, 1982). For a dissenting view, see Arthur Fine, *The Shaky Game: Einstein, realism, and quantum theory* (Chicago: University of Chicago Press, 1986).

42. See especially Popper, *Conjectures and Refutations* (New York: Harper & Row, 1963) and *The Logic of Scientific Discovery* (London: Hutchinson, 1959).

43. A. Einstein, B. Podolsky and N. Rosen, 'Can Quantum-Mechanical Description of Reality be Considered Complete?', *Physical Review*, Series 2, Vol. 47 (1935), pp. 777–80.

44. For further discussion see Brown, *Smoke and Mirrors* (op. cit.).

45. Popper, *Quantum Theory and the Schism in Physics* (op. cit.), p. 172.

46. On the 'hidden variables' theory, see also David Bohm, *Quantum Theory* (Englewood Cliffs, NJ: Prentice-Hall, 1951); David Bohm and Basil J. Hiley, *The Undivided Universe: An ontological interpretation of quantum theory* (London: Routledge, 1993).

47. See Niels Bohr, *Atomic Theory and the Description of Nature* (Cambridge: Cambridge University Press, 1934) and *Atomic Physics and Human Knowledge* (New York: Wiley, 1958); also Henry J. Folse, *The Philosophy of Niels Bohr: The framework of complementarity* (Amsterdam: North-Holland, 1985); John Honner, *The Description of Nature: Niels Bohr and the philosophy of quantum physics* (Oxford: Clarendon Press, 1987); Dugald Murdoch, *Niels Bohr's Philosophy of Physics* (Cambridge: Cambridge University Press, 1987).

48. Popper, *Quantum Theory and the Schism in Physics* (op. cit.), pp. 9–10.

49. See Werner Heisenberg, *Philosophical Problems of Quantum Physics*, trans. F.C. Hayes (Woodbridge, Conn.: Ox Bow Press, 1979); William C. Price and Seymour S. Chissick (eds), *The Uncertainty Principle and the Foundation of Quantum Mechanics* (New York: Wiley, 1977).

50. Popper, *Quantum Theory and the Schism in Physics* (op. cit.), p. 9.

51. See for instance – from a range of (inter)disciplinary perspectives – Jean-François Lyotard, *The Postmodern Condition: A report on knowledge*, trans. Geoff Bennington and Brian Massumi (Manchester:

Manchester University Press, 1984); Andrew Pickering, *Constructing Quarks: A sociological history of particle physics* (Edinburgh: Edinburgh University Press, 1983); Arkady Plotnitsky, *Complementarity: Anti-epistemology after Bohr and Derrida* (Durham, NC & London: Duke University Press, 1994); also – for a strongly dissenting view – Roger Trigg, *Reality at Risk: A defence of realism in philosophy and the sciences* (Brighton: Harvester, 1980).

52. Kuhn, *The Structure of Scientific Revolutions* (op. cit.).
53. Jacques Derrida, 'The Supplement of Copula: Philosophy *before* linguistics', in *Margins of Philosophy*, trans. Alan Bass (Chicago: University of Chicago Press, 1982), pp. 175–205.
54. Emile Benveniste, *Problems in General Linguistics*, trans. Mary E. Meek (Coral Gables: University of Miami Press, 1971).
55. David Papineau, *Reality and Representation* (Oxford: Blackwell, 1987), pp. 24–5.
56. Christopher Norris, 'On Not Going Relativist (where it counts)', in *The Contest of Faculties: Philosophy and theory after deconstruction* (London: Methuen, 1985); also Norris, 'Truth, Meaning and Right Interpretation' (op. cit.).
57. See Davidson, 'On the Very Idea of a Conceptual Scheme' (op. cit.); also Richard Grandy, 'Reference, Meaning, and Belief', *Journal of Philosophy*, Vol. 70 (1973), pp. 439–52; J.E. Malpas, 'The Nature of Interpretive Charity', *Dialectica*, Vol. 42 (1988), pp. 17–36; Bruce Vermazen, 'General Beliefs and the Principle of Charity', *Philosophical Studies*, Vol. 42 (1982), pp. 111–18.
58. For a discussion of recent debate on this topic see R. Shope, *The Analysis of Knowing* (Princeton, NJ: Princeton University Press, 1983).
59. See Rorty, 'Is Truth a Goal of Enquiry?' (op. cit.).
60. See Paul K. Feyerabend, *Against Method* (London: New Left Books, 1975) and – albeit from a very different philosophical standpoint – the essays collected in Michael Dummett, *Truth and Other Enigmas* (London: Duckworth, 1978).
61. See for instance Larry Laudan, *Progress and Its Problems* (Berkeley & Los Angeles: University of California Press, 1977); Peter Lipton, *Inference to the Best Explanation* (London: Routledge, 1993); Nicholas Rescher, *Scientific Progress* (Oxford: Blackwell, 1979); Peter J. Smith, *Realism and the Progress of Science* (Cambridge: Cambridge University Press, 1981).
62. See Feyerabend, *Against Method* (op. cit.) and *Science in a Free Society* (London: New Left Books, 1978); Rorty, *Consequences of Pragmatism* (op. cit.).
63. See for instance Martin Hollis and Steven Lukes (eds), *Rationality and Relativism* (Oxford: Blackwell, 1982); Michael Krausz (ed.), *Relativism: Interpretation and confrontation* (Notre Dame, Ind.: University of Notre Dame Press, 1989); Larry Laudan, *Science and Relativism: Some key controversies in the philosophy of science* (Chicago: University of Chicago Press, 1990); W.H. Newton-Smith, *The Rationality of Science* (op. cit.).
64. Donald Davidson, *Essays on Actions and Events* (Oxford: Clarendon Press, 1980).

65. Rorty, 'Pragmatism, Davidson, and Truth', in *Objectivity, Relativism, and Truth* (Cambridge: Cambridge University Press, 1991), pp. 126–150. On this topic, see also Colin McGinn, 'Charity, Interpretation and Belief', *Journal of Philosophy*, Vol. 74 (1977), pp. 521–35; Rorty, 'Transcendental Arguments, Self-Reference, and Pragmatism', in P. Bieri, R.-P. Horstmann and L Krüger (eds), *Transcendental Arguments and Science* (Dordrecht: D. Reidel, 1979), pp. 95–9: Barry Stroud, 'Transcendental Arguments', *Journal of Philosophy*, Vol. 65 (1968), pp. 242–56.

66. This debate is well summarized in Kirkham, *Theories of Truth* (note 6, above).

67. Alfred Tarski, 'The Semantic Conception of Truth', in Leonard Linsky (ed.), *Semantics and the Philosophy of Language* (Urbana, Ill.: University of Illinois Press, 1952), p. 34.

68. Kirkham, *Theories of Truth* (op. cit.).

69. Ibid., p. 196.

70. Ibid., pp. 191–2.

71. Ibid., p. 193.

72. Davidson, 'The Structure and Content of Truth' (op. cit.).

73. Kirkham, *Theories of Truth* (op. cit.), p. 193.

74. See Saul Kripke, *Naming and Necessity* (Oxford: Blackwell, 1980) and the essays by Donnellan, Evans, Kripke, Putnam and others collected in Stephen Schwartz (ed.), *Naming, Necessity, and Natural Kinds* (Ithaca, NY: Cornell University Press, 1977); also David Wiggins, *Sameness and Substance* (Oxford: Blackwell, 1980). For related (causal-realist) arguments in epistemology, ontology, and philosophy of science, see especially D.M. Armstrong, *What Is a Law of Nature?* (Cambridge: Cambridge University Press, 1983); Roy Bhaskar, *Scientific Realism and Human Emancipation* (London: Verso, 1986) and *Reclaiming Reality: a critical introduction to contemporary philosophy* (Verso, 1989); Rom Harré and E.H. Madden, *Causal Powers* (Blackwell, 1975); Wesley C. Salmon, *Scientific Explanation and the Causal Structure of the World* (Princeton, NJ: Princeton University Press, 1984) and *Four Decades of Scientific Explanation* (Minneapolis: University of Minnesota Press, 1989); Brian Skyrms, *Causal Necessity* (New Haven, Conn.: Yale University Press, 1980); Michael Tooley, *Causation: A realist approach* (Blackwell, 1988).

75. See Heidegger, *The Question Concerning Technology and Other Essays*, trans. William Lovitt (New York: Harper & Row, 1977).

76. See note 35, above.

77. See note 74, above.

78. Hilary Putnam, *The Many Faces of Realism* (La Salle, Ill.: Open Court, 1987); *Realism with a Human Face* (Cambridge, Mass.: Harvard University Press, 1990).

2. Complex Words versus Minimalist Semantics

1. William Empson, *The Structure of Complex Words* (2nd edn, revised, London: Chatto & Windus, 1969). All further references given by title and page number in the text.

2. See for instance the passing references to Empson in Davidson's essay 'What Metaphors Mean', *Inquiries into Truth and Interpretation* (Oxford: Clarendon Press, 1984), pp. 245–64.

3. See especially Donald Davidson, 'A Nice Derangement of Epitaphs', in R. Grandy and R. Warner (eds), *Philosophical Grounds of Rationality: Intentions, categories, ends* (London: Oxford University Press, 1986), pp. 157–74.

4. See for instance the essays collected in W.J.T. Mitchell (ed.), *Against Theory: Literary theory and the new pragmatism* (Chicago: University of Chicago Press, 1985); also S. Pradhan, 'Minimalist Semantics: Davidson and Derrida on meaning, use, and convention', *Diacritics*, Vol. 16 (Spring 1986), pp. 66–77 and Samuel C. Wheeler, 'Indeterminacy of French Translation: Derrida and Davidson', in Ernest Lepore (ed.), *Truth and Interpretation: Perspectives on the philosophy of Donald Davidson* (Oxford: Blackwell, 1986).

5. Empson, *Seven Types of Ambiguity* (2nd edn, revised, Harmondsworth: Penguin, 1961).

6. Empson, 'Sense in *The Prelude*', *Complex Words* (op. cit.), pp. 289–305.

7. See for instance Noam Chomsky, *Current Issues in Linguistic Theory* (The Hague: Mouton, 1966); *Studies on Semantics in Generative Grammar* (Mouton, 1966); *Language and Problems of Knowledge* (Cambridge, Mass.: MIT Press, 1988); Jerry Fodor, *Representations* (MIT Press, 1981) and *The Modularity of Mind* (MIT Press, 1983).

8. Christopher Norris, *William Empson and the Philosophy of Literary Criticism* (London: Athlone Press, 1978).

9. On this distinction between 'prior' and 'passing' theories, see Davidson, 'A Nice Derangement of Epitaphs' (op. cit.).

10. See especially Davidson, 'On the Very Idea of a Conceptual Scheme', in *Inquiries into Truth and Interpretation* (op. cit.), pp. 183–98; also Richard Grandy, 'Reference, Meaning and Belief', *Journal of Philosophy*, Vol. 70 (1973), pp. 439–52; Colin McGinn, 'Charity, Interpretation and Belief', *Journal of Philosophy*, Vol. 74 (1977), pp. 521–35; Bjorn Ramberg, *Donald Davidson's Philosophy of Language: An introduction* (Oxford: Blackwell, 1989).

11. On the subject of 'radical translation', see W.V. Quine, *Word and Object* (Cambridge, Mass.: MIT Press, 1960); also Quine, *Ontological Relativity and Other Essays* (New York: Random House, 1966); Davidson, 'On the Very Idea of a Conceptual Scheme' (op. cit.); David Lewis, 'Radical Translation', *Philosophical Papers*, Vol. 1 (London: Oxford University Press, 1983); Colin McGinn, 'Radical Interpretation and Epistemology', in Lepore (ed.), *Truth and Interpretation* (op. cit.), pp. 356–68.

12. W.H. Newton-Smith, *The Rationality of Science* (London: Routledge & Kegan Paul, 1981), p. 163.

13. Ibid., p. 163.

14. See for instance Richard Rorty, 'The World Well Lost', in *Consequences of Pragmatism* (Brighton: Harvester, 1982), pp. 3–18; 'Pragmatism, Davidson, and Truth', in *Objectivity, Relativism, and Truth* (Cam-

bridge: Cambridge University Press, 1991), pp. 126–50; 'Is Truth a Goal of Enquiry? Davidson *versus* Wright', *Philosophical Quarterly*, Vol. 45 (1995), pp. 281–300.

15. I.A. Richards, *Principles of Literary Criticism* (London: Paul Trench Trubner, 1924); Cleanth Brooks, *The Well Wrought Urn: Studies in the structure of poetry* (New York: Harcourt Brace, 1947).

16: Davidson, 'A Nice Derangement of Epitaphs' (op. cit.). His point, once again, is that we interpret malapropisms – along with metaphors, novel turns of phrase, Freudian slips, idiomatic usages, *hapax legomena*, etc. – by application of the standard 'Principle of Charity' plus 'wit, luck and wisdom' eked out (where need be) by contextual cues and clues. Thus for Davidson 'there is no such thing as a language', at any rate not in the sense of that term understood by most linguists, philosophers, and theorists of interpretation.

17. Leonard Bloomfield, *Language* (London: Allen & Unwin, 1935).

18. Empson, *Seven Types of Ambiguity* (op. cit.).

19. See for instance Jean-François Lyotard, *The Postmodern Condition: A report on knowledge*, trans. Geoff Bennington and Brian Massumi (Manchester: Manchester University Press, 1984).

20. See Stanley Fish, *Is There a Text in This Class? The authority of interpretive communities* (Cambridge, Mass.: Harvard University Press, 1983) and *Doing What Comes Naturally: Change, rhetoric, and the practice of theory in literary and legal studies* (Oxford: Clarendon Press, 1989); also Rorty, *Consequences of Pragmatism* (op. cit.) and *Contingency, Irony, and Solidarity* (Cambridge: Cambridge University Press, 1989).

21. For a strongly argued critique of these notions, see Karl R. Popper, *Quantum Theory and the Schism in Physics* (London: Hutchinson, 1982).

22. See the essays collected in A.J. Ayer (ed.), *Logical Positivism* (New York: Free Press, 1953); also Friedrich Waismann, 'Verifiability' and Isaiah Berlin, 'Verification', in G.H.R. Parkinson (ed.), *The Theory of Meaning* (London: Oxford University Press, 1976), pp. 15–34 & 35–60.

23. See especially Quine, 'Two Dogmas of Empiricism', *From a Logical Point of View* (Cambridge, Mass.: Harvard University Press, 1953), pp. 20–46.

24. Richards, *Principles of Literary Criticism* (op. cit.).

25. David Papineau, *Theory and Meaning* (London: Oxford University Press, 1979); *Reality and Representation* (Oxford: Blackwell, 1987); *Philosophical Naturalism* (Blackwell, 1993).

26. Alfred Tarski, 'The Concept of Truth in Formalized Languages', in *Logic, Semantics and Metamathematics*, trans. J.H. Woodger (London: Oxford University Press, 1956), pp. 152–278; also Davidson, 'In Defence of Convention T' and 'The Method of Truth in Metaphysics', *Inquiries into Truth and Interpretation* (op. cit.), pp. 65–76 & 199–214.

27. Davidson, 'On the Very Idea of a Conceptual Scheme' (op. cit.).

28. Papineau, *Reality and Representation* (op. cit.), p. 99. See also Richard Grandy, 'Reference, Meaning and Belief', *Journal of Philosophy*, Vol. 70 (1973), pp. 439–52. So far as I know Grandy was the first to explicate this distinction between 'charity' and 'humanity' as principles of interpretation.

29. See Michael Dummett, *Truth and Other Enigmas* (London: Duckworth, 1978).
30. Papineau, *Reality and Representation* (op. cit.), p. 87.
31. See also Robert L. Arrington, *Rationalism, Realism, and Relativism: Perspectives in contemporary moral epistemology* (Ithaca, NY: Cornell University Press, 1989); Rom Harré, *Varieties of Realism: A rationale for the social sciences* (Oxford: Blackwell, 1986); Martin Hollis and Steven Lukes (eds), *Rationality and Relativism* (Blackwell, 1982).
32. Papineau, *Reality and Representation* (op. cit.), p. 124.
33. Ibid., p. 125.
34. Rorty, 'Transcendental Arguments, Self-Reference, and Pragmatism', in P. Bieri, R. Horstmann and L. Krüger (eds), *Transcendental Arguments and Science* (Dordrecht: D. Reidel, 1979), pp. 77–103; p. 78; also Mark Okrent, 'The Metaphilosophical Consequences of Pragmatism', in Avner Cohen and Marcelo Dascal (eds), *The Institution of Philosophy: A discipline in crisis?* (La Salle, Ill.: Open Court, 1989), pp. 177–98.
35. Davidson, *Essays on Actions and Events* (Oxford: Clarendon Press, 1980).
36. For further discussion, see the essays collected in Ernest LePore and Brian McLaughlin (eds), *Actions and Events: Perspectives on the philosophy of Donald Davidson* (Oxford: Blackwell, 1985).
37. Papineau, *Reality and Representation* (op. cit.), p. 124.
38. Paul Ricoeur, *Oneself as Another*, trans. Kathleen Blamey (Chicago: University of Chicago Press, 1992), p. 74.
39. Gilbert Ryle, *On Thinking* (Oxford: Blackwell, 1979) and *Aspects of Mind*, ed. René Meyer (Blackwell, 1993).
40. Ricoeur, *Oneself as Another* (op. cit.), p. 75.
41. Davidson, *Inquiries into Truth and Interpretation* (op. cit.), p. 279.
42. Davidson, 'A Nice Derangement of Epitaphs'(op. cit.), p. 173.
43. Ibid., p. 164.
44. Ibid., p. 161.
45. Davidson, 'Communication and Convention', in *Inquiries into Truth and Interpretation* (op. cit.), pp. 265–80; p. 280.
46. Papineau, *Reality and Representation* (op. cit.).
47. See Empson, 'Sense and Sensibility', 'Sense in "Measure for Measure" ', 'Sense in *The Prelude*', and 'Sensible and Candid', in *The Structure of Complex Words* (op. cit.), pp. 250–310.
48. Empson, *Seven Types of Ambiguity* (op. cit.), pp. 151–4.
49. C.L. Stevenson, *Ethics and Language* (New Haven, Conn.: Yale University Press, 1944); Cleanth Brooks, *The Well Wrought Urn* (op. cit.); W.K. Wimsatt, *The Verbal Icon: Studies in the meaning of poetry* (Lexington, Ky: University of Kentucky Press, 1954).
50. Brooks, *The Well Wrought Urn* (op. cit.), p. 138.
51. Wimsatt, *The Verbal Icon* (op. cit.).
52. See for instance John Crowe Ransom, 'Mr. Empson's Muddles', *Southern Review*, Vol. 4 (1938/9), pp. 322–39.
53. Empson, *Seven Types of Ambiguity* (op. cit.), pp. 224–6 and 226–33.
54. Ibid., p. 195.

55. Ibid., p. 197.
56. See the articles and reviews collected in Empson, *Argufying: Essays on literature and culture*, ed. John Haffenden (London: Chatto & Windus, 1987); especially Section I, 'Literary Interpretation: The language machine', pp. 67–189.
57. Empson, 'Herbert's Quaintness', in *Argufying* (op. cit.), pp. 256–9; p. 257.
58. Empson, 'Thy Darling in an Urn', in *Argufying* (op. cit.), 282–8; p. 283.
59. Empson, 'Still the Strange Necessity', in *Argufying* (op. cit.), pp. 120–28; p. 124.
60. Ibid., p. 126.
61. Empson, *Seven Types of Ambiguity* (op. cit.), pp. 196–7.

3. Doubling Castle or the Slough of Despond

1. Richard Rorty (ed.), *The Linguistic Turn* (Chicago: University of Chicago Press, 1967).
2. Gottlob Frege, 'On Sense and Reference', in P. Geach and M. Black (eds), *Translations from the Philosophical Writings of Gottlob Frege* (Oxford: Blackwell, 1952), pp. 56–78; Bertrand Russell, 'On Denoting', *Mind*, Vol. 14 (1905), pp. 479–93.
3. J.L. Austin, 'A Plea for Excuses', in *Philosophical Papers* (Oxford: Clarendon Press, 1961), pp. 175–204; p. 182.
4. See Rorty, *Philosophy and the Mirror of Nature* (Oxford: Blackwell, 1980); *Consequences of Pragmatism* (Brighton: Harvester, 1982); *Contingency, Irony, and Solidarity* (Cambridge: Cambridge University Press, 1989).
5. Donald Davidson, 'On the Very Idea of a Conceptual Scheme', in *Inquiries into Truth and Interpretation* (Oxford: Clarendon Press, 1984), pp. 183–98 and 'A Nice Derangement of Epitaphs', in R. Grandy and R. Warner (eds), *Philosophical Grounds of Rationality: Intentions, categories, ends* (London: Oxford University Press, 1986), pp. 157–74.
6. See for instance Rorty, 'Pragmatism, Davidson, and Truth', in Ernest Lepore (ed.), *Truth and Interpretation: Perspectives on the philosophy of Donald Davidson* (Oxford: Blackwell, 1986), pp. 333–55; also Rorty, 'Is Truth a Goal of Enquiry? Davidson *versus* Wright', *Philosophical Quarterly*, Vol. 45 (1995), pp. 281–300.
7. Davidson, 'A Nice Derangement of Epitaphs' (op. cit.).
8. Davidson, 'On the Very Idea of a Conceptual Scheme' (op. cit.); also 'Truth and Meaning', 'Radical Interpretation' and 'The Method of Truth in Metaphysics', in *Inquiries into Truth and Interpretation* (op. cit.), pp. 17–36, 65–76 and 199–214.
9. See Alfred Tarski, 'The Concept of Truth in Formalized Languages', *Logic, Semantics and Metamathematics*, trans. J.H. Woodger (London: Oxford University Press, 1956), pp. 152–278.
10. Stephen Schiffer, *Remnants of Meaning* (Cambridge, Mass.: MIT Press, 1987), p. 179.
11. Ibid., pp. 179–80.
12. See note 8, above.

13. Rorty, 'Transcendental Arguments, Self-Reference, and Pragmatism', in P. Bieri, R. Horstmann, and L. Krüger (eds), *Transcendental Arguments and Science* (Dordrecht: D. Reidel, 1979), p. 78.

14. See Tarski, 'The Concept of Truth in Formalized Languages' (op. cit.) and Davidson, 'In Defence of Convention T', in *Inquiries into Truth and Interpretation* (op. cit.), pp. 65-76.

15. Compare for instance the realist position adopted in Hilary Putnam, *Philosophical Papers*, Vols 1 & 2 (Cambridge: Cambridge University Press, 1975) with the scaled-down 'internal' variant proposed in *The Many Faces of Realism* (La Salle, Ill.: Open Court, 1987), *Realism with a Human Face* (Cambridge, Mass.: Harvard University Press, 1990), and *Renewing Philosophy* (Harvard UP, 1992). See also the various responses to his work in Peter Clark and Bob Hale (eds), *Reading Putnam* (Oxford: Blackwell, 1993).

16. Schiffer, *Remnants of Meaning* (op. cit.). All further references given by '*Remnants*' and page-number in the text. See also Schiffer, *Meaning* (Oxford: Clarendon Press, 1972).

17. Jerry A. Fodor, *A Theory of Content and Other Essays* (Cambridge, Mass.: MIT Press, 1992), p. 179.

18. See Fodor, *A Theory of Content* (op. cit.); also *Representations* (Cambridge, Mass.: MIT Press, 1981); *The Modularity of Mind* (MIT Press, 1983); *Psychosemantics: The problem of meaning in the philosophy of mind* (MIT 1987); *The Elm and the Expert: mentalese and its semantics* (MIT, 1994).

19. Thus Quine: 'If we are limning the true and ultimate structure of reality, the canonical scheme for us is the austere scheme that knows no quotation but direct quotation and no propositional attitudes but only the physical constitution and behavior of organisms'. (W.V. Quine, *Word and Object* [Cambridge, Mass.: MIT Press, 1960], p. 221.)

20. Davidson, 'A Nice Derangement of Epitaphs' (op. cit.).

21. Fodor, *A Theory of Content* (op. cit.), p. 4.

22. Ludwig Wittgenstein, *Philosophical Investigations*, trans. G.E.M. Anscombe (2nd edn., Oxford: Blackwell, 1958).

23. See especially Peter Winch, *The Idea of a Social Science and Its Relation to Philosophy* (London: Routledge & Kegan Paul 1958); also David Bloor, *Wittgenstein: A social theory of knowledge* (New York: Columbia University Press, 1983).

24. See David Papineau, *Reality and Representation* (Oxford: Blackwell, 1987); also *Theory and Meaning* (London: Oxford University Press, 1979) and *Philosophical Naturalism* (Blackwell, 1993).

25. Davidson, 'A Nice Derangement of Epitaphs' (op. cit.).

26. Papineau, *Reality and Representation* (op. cit.).

27. W.V. Quine, *Word and Object* (Cambridge, Mass.: MIT Press, 1960); also Quine, *From a Logical Point of View* (Cambridge, Mass.: Harvard University Press, 1953); *Ontological Relativity and Other Essays* (New York: Columbia University Press, 1969).

28. Davidson, 'On the Very Idea of a Conceptual Scheme' (op. cit.).

29. Papineau, *Reality and Representation* (op. cit.), p. 98.

30. Ibid., p. 98.
31. Ibid., p. 99.
32. Ibid., p. 99.
33. See Winch, *The Idea of a Social Science* and Bloor, *Wittgenstein: A social theory of knowledge* (note 23, above); also Bloor, *Knowledge and Social Imagery* (London: Routledge & Kegan Paul, 1976); Peter L. Berger and Thomas Luckmann, *The Social Construction of Reality: A treatise on the sociology of knowledge* (Harmondsworth: Penguin, 1967); Steve Fuller, *Social Epistemology* (Bloomington, Ind.: Indiana University Press, 1988); Clifford Geertz, *The Interpretation of Cultures* (New York: Basic Books, 1973) and *Local Knowledge: Further essays on interpretive authority* (Basic Books, 1983); Derek L. Phillips, *Wittgenstein and Scientific Knowledge: A sociological perspective* (London: Macmillan, 1977); Michael Walzer, *Interpretation and Social Criticism* (Cambridge, Mass.: Harvard University Press, 1987).
34. See for instance – from various standpoints – Daniel Bell, *Communitarianism and Its Critics* (Oxford: Clarendon Press, 1993); David O. Brink, *Moral Realism and the Foundations of Ethics* (Cambridge: Cambridge University Press, 1989); Jürgen Habermas, *Justification and Application: Remarks on discourse ethics*, trans. Ciaran P. Cronin (Cambridge: Polity Press, 1993); Sabina Lovibond, *Realism and Imagination in Ethics* (Oxford: Blackwell, 1983); Onora O'Neill, - *Constructions of Reason: Explorations of Kant's practical philosophy* (Cambridge UP, 1989); David Rasmussen, *Universalism versus Communitarianism: Contemporary debates in ethics* (Cambridge, Mass.: MIT Press, 1990).
35. O'Neill, *Constructions of Reason* (op. cit.).
36. See Winch, *The Idea of a Social Science* (op. cit.); also *Ethics and Action* (London: Routledge & Kegan Paul, 1972).
37. Alasdair MacIntyre, *Against the Self-Images of the Age* (London: Duckworth, 1971).
38. MacIntyre, *After Virtue: A study in moral theory* (London: Duckworth, 1981); *Whose Justice? Which Rationality?* (Duckworth, 1988).
39. O'Neill, *Constructions of Reason* (op. cit.), p. 172.
40. Ibid., p. 175.
41. See especially Saul Kripke, *Naming and Necessity* (Oxford: Blackwell, 1980); Stephen P. Schwartz (ed.), *Naming, Necessity, and Natural Kinds* (Ithaca, NY: Cornell University Press, 1977); David Wiggins, *Sameness and Substance* (Blackwell, 1980).
42. See Frege, 'On Sense and Reference' (op. cit.).
43. See Kripke, *Naming and Necessity* (op. cit.); also Putnam, 'Is Semantics Possible?' and 'Meaning and Reference', in Schwartz (ed.), *Naming, Necessity, and Natural Kinds* (op. cit.), pp. 102–18 and 119–32.
44. Gareth Evans, 'The Causal Theory of Names', in Schwartz, *Naming, Necessity, and Natural Kinds* (op. cit.), pp. 192–215.
45. Ibid., pp. 200–1.
46. Ibid., p. 201.
47. Ibid., p. 204.
48. Ibid., p. 204.

49. Recent interest in these topics is largely owing to E.L. Gettier, 'Is Justified True Belief Knowledge?', *Analysis*, Vol. 23 (1963). For some ingenious (if often rather contrived) examples, see R. Shope, *The Analysis of Knowing* (Princeton, NJ: Princeton University Press, 1983).
50. Evans, 'The Causal Theory of Names' (op. cit.), p. 214.
51. Ibid., p. 214.
52. Ibid., pp. 214–15.
53. See note 41, above; also Roy Bhaskar, *Scientific Realism and Human Emancipation* (London: Verso, 1986); Rom Harré, *The Principles of Scientific Thinking* (Chicago: University of Chicago Press, 1970) and *Varieties of Realism: A rationale for the social sciences* (Oxford: Blackwell, 1986); Rom Harré and E.H. Madden, *Causal Powers* (Blackwell, 1975); Jarrett Leplin (ed.), *Scientific Realism* (Berkeley & Los Angeles: University of California Press, 1984); Mark Platts (ed.), *Reference, Truth and Reality: Essays on the philosophy of language* (London: Routledge & Kegan Paul, 1980); Hilary Putnam, *Realism and Reason* (Cambridge: Cambridge University Press, 1993); Wesley C. Salmon, *Scientific Explanation and the Causal Structure of the World* (Princeton, NJ: Princeton University Press, 1984) and *Four Decades of Scientific Explanation* (Minneapolis: University of Minnesota Press, 1989); J.J.C. Smart, *What Is a Law of Nature?* (Cambridge UP, 1983); Peter J. Smith, *Realism and the Progress of Science* (Cambridge UP, 1981); M. Tooley, *Causation: A realist approach* (Oxford: Blackwell, 1988).
54. For some shrewd suggestions as to just why this should be the case, see J. L. Mackie, *Logic and Knowledge: Selected papers, Vol. 1* (Oxford: Clarendon Press, 1985).

4. Complex Words, Natural Kinds, and the Justification of Belief

1. See especially Donald Davidson, 'A Nice Derangement of Epitaphs', in R. Grandy and R. Warner (eds), *Philosophical Grounds of Rationality: Intentions, categories, ends* (London: Oxford University Press, 1986), pp. 157–74.
2. On the Principle of Charity see Davidson, *Inquiries into Truth and Interpretation* (Oxford: Clarendon Press, 1974); also Richard Grandy, 'Reference, Meaning and Belief', *Journal of Philosophy*, Vol. 70 (1973), pp. 439–52; J.E. Malpas, 'The Nature of Interpretative Charity', *Dialectica*, Vol. 42 (1988), pp. 17–36; Colin McGinn, 'Charity, Interpretation and Belief', *Journal of Philosophy*, Vol. 74 (1977), pp. 521–35; David Papineau, *Reality and Representation* (Oxford: Blackwell, 1987); Michael Williams, 'Scepticism and Charity', *Ratio* (New Series), Vol. 1 (1988), pp. 176–94.
3. Stephen Schiffer, *Fragments of Meaning* (Cambridge, Mass.: MIT Press, 1987).
4. For this earlier, more confident analytical approach see Schiffer, *Meaning* (London: Oxford University Press, 1972).
5. See Schiffer, *Remnants of Meaning* (op. cit.), pp. 49–71.
6. Davidson, 'A Nice Derangement of Epitaphs' (op. cit.).

7. Ibid.

8. Ibid., p. 170.

9. William Empson, *The Structure of Complex Words* (2nd edn, London: Chatto & Windus, 1969).

10. Empson, *Complex Words*, p. 174. See also Davidson, 'What Metaphors Mean', in *Inquiries into Truth and Interpretation* (op. cit.), pp. 245–64.

11. Empson, 'The English Dog', in *Complex Words* (op. cit.), pp. 158–74.

12. Elliott Sober, *Philosophy of Biology* (London: Oxford University Press, 1993), p. 144. See also D. Hull, *Philosophy of Biological Sciences* (Englewood Cliffs, NJ: Prentice-Hall, 1974); M. Ruse, *Philosophy of Biology Today* (Albany, NY: State University of New York Press, 1988).

13. Sober, *Philosophy of Biology* (op. cit.), p. 153.

14. Ibid., p. 144. For a sample of Foucault's ultra-nominalist approach, see his Preface to *The Order of Things* (New York: Random House, 1973), pp. xv–xxiv; also – in a less extravagant but somewhat related vein – John Dupré, *The Disorder of Things: Metaphysical foundations of the disunity of science* (Cambridge, Mass.: Harvard University Press, 1993).

15. Sober, *Philosophy of Biology* (op. cit.), p. 144. See also W.V. Quine, 'Two Dogmas of Empiricism', in *From a Logical Point of View* (Cambridge, Mass.: Harvard University Press, 1953), pp. 20–46 and *Ontological Relativity and Other Essays* (New York: Columbia University Press, 1969).

16. William Empson, *Seven Types of Ambiguity* (2nd edn, Harmondsworth: Penguin, 1961), p. 1.

17. Ibid., p. 1.

18. See note 15, above.

19. Hilary Kornblith, *Inductive Inference and Its Natural Ground: An essay in naturalistic epistemology* (Cambridge, Mass.: MIT Press, 1993), p. 15. See also Kornblith (ed.), *Naturalizing Epistemology* (MIT Press, 1985).

20. Foucault cites Borges, who himself makes (fictive) reference to 'a certain Chinese encyclopaedia' wherein animals are classified as follows: '(a) belonging to the Emperor, (b) embalmed, (c) tame, (d) sucking pigs, (e) sirens, (f) fabulous, (g) stray dogs, (h) included in the present classification, (i) frenzied, (j) innumerable, (k) drawn with a very fine camelhair brush, (l) *etcetera*, (m) having just broken the water pitcher, (n) that from a long way off look like flies'. (Foucault, *The Order of Things* [op. cit.], p. xv)

21. Kornblith, *Inductive Inference and Its Natural Ground* (op. cit.), p. 37.

22. Ibid., p. 36.

23. See Hilary Putnam, 'Is Semantics Possible?' and 'Meaning and Reference', in Stephen P. Schwartz (ed.), *Naming, Necessity, and Natural Kinds* (Ithaca, NY: Cornell University Press, 1977), pp. 102–18 and 119–32.

24. Saul Kripke, *Naming and Necessity* (Oxford: Blackwell, 1980).

25. See for instance Gareth Evans, 'The Causal Theory of Names', in Schwartz (ed.), *Naming, Necessity, and Natural Kinds* (op. cit.), pp. 192–215.

26. Empson, *The Structure of Complex Words* (op. cit.), p. 163. All further references given by title and page-number in the text.
27. Empson, 'Timon's Dog', in *Complex Words*, pp. 175–84.
28. See David Papineau, *Reality and Representation* (cited in note 2, above); also Papineau, *For Science in the Social Sciences* (London: Macmillan, 1978), *Theory and Meaning* (London: Oxford University Press, 1979), and *Philosophical Naturalism* (Oxford: Blackwell, 1993).
29. See especially Papineau, *Reality and Representation* (op. cit.).
30. Empson, 'Sense and Sensibility', 'Sense in *Measure for Measure*', 'Sense in *The Prelude*', and 'Sensible and Candid', in *The Structure of Complex Words* (op. cit.), pp. 250–310.
31. Davidson, 'A Nice Derangement of Epitaphs' (op. cit.).
32. Thus Empson: 'What he [Orwell] calls "doublethink", a process of intentional but genuine self-deception, easy to reach but hard to hold permanently, really does seem a positive capacity of the human mind, so curious and so important in its effects that any theory in this field needs to reckon with it' (*Complex Words*, p. 83n).
33. Davidson, 'A Nice Derangement of Epitaphs' (op. cit.).
34. Schiffer, *Remnants of Meaning* (op. cit.).
35. Papineau, *Reality and Representation* (op. cit.).
36. See the various articles and reviews collected in William Empson, *Argufying: Essays on literature and culture*, ed. John Haffenden (London: Chatto & Windus, 1987), especially Section 1, 'Literary Interpretation: the language machine', pp. 67–189.
37. See for instance W.K. Wimsatt, *The Verbal Icon: Studies in the meaning of poetry* (Lexington: University of Kentucky Press, 1954) and Cleanth Brooks, *The Well Wrought Urn: Studies in the structure of poetry* (New York: Harcourt Brace, 1947).
38. Cited by Empson, *Complex Words*, p. 173.
39. Paul Fussell, *The Rhetorical World of Augustan Humanism* (Oxford: Clarendon Press, 1965).
40. Empson, *Seven Types of Ambiguity* (op. cit.), pp. 243–56.
41. Empson, 'Milton and Bentley', in *Some Versions of Pastoral* (Harmondsworth: Penguin, 1966), pp. 121–55.
42. Ibid., p. 126.
43. Empson, *Milton's God* (2nd edn., revised, London: Chatto & Windus, 1965).
44. Empson, 'Milton and Bentley' (op. cit.), p. 125.
45. Empson, 'All in *Paradise Lost*', in *The Structure of Complex Words* (op. cit.), pp. 101–4.
46. Samuel Johnson, 'Preface to Shakespeare [1765]', in W.K. Wimsatt (ed.), *Johnson on Shakespeare* (Harmondsworth: Penguin, 1965); *Lives of the English Poets* (2 vols, London: Dent, undated).
47. In this connection see Paul H. Fry, *The Reach of Criticism* (New Haven, Conn.: Yale University Press, 1983).
48. Schiffer, *Remnants of Meaning* and Davidson, 'A Nice Derangement of Epitaphs' (cited above).

49. See for instance Kathleen Williams, *Jonathan Swift and the Age of Compromise* (Kansas: University of Kansas Press, 1958).

50. Empson, *Seven Types of Ambiguity* (op. cit.), p. 252.

51. Ibid., p. 252.

52. Ibid., p. 254.

53. Ibid., p. 252.

54. See W.V. Quine, *Word and Object* (Cambridge, Mass.: MIT Press, 1960); also references in note 15, above.

55. Davidson, 'A Nice Derangement of Epitaphs' (op. cit.).

56. Schiffer, *Remnants of Meaning* (op. cit.), p. 70.

57. See W.K. Wimsatt, *The Verbal Icon* (op. cit.), in particular his essay 'The Intentional Fallacy', co-authored with Monroe K. Beardsley. For Empson's counter-arguments attacking what he calls the 'Wimsatt Law' and defending the appeal to authorial intention, see especially his *Using Biography* (London: Chatto & Windus, 1984) and *Essays on Shakespeare*, David B. Pirie (ed.) (Cambridge: Cambridge University Press, 1986). This issue is also taken up at various points – most often *contra* Wimsatt and other proponents of the anti-intentionalist line – in the articles and reviews collected in Empson, *Argufying* (op. cit.). See especially 'The Verbal Analysis' (pp. 104–9), 'Still the Strange Necessity' (120–8), 'Rhythm and Imagery in English Poetry' (147–66), 'Thy Darling in an Urn' (282–8), 'The Love of Definition' (264–72), and 'The Ancient Mariner' (297–319).

58. See Wimsatt and Beardsley, 'The Intentional Fallacy' (op. cit.).

59. See Brooks, *The Well Wrought Urn* and Wimsatt, *The Verbal Icon* (note 37, above).

60. Empson, *Argufying* (op. cit.), pp. 124–5.

61. Ibid., p. 165.

62. See Empson, *Milton's God* (op.cit.) and *Faustus and the Censor*, John Henry Jones (ed.) (Oxford: Blackwell, 1987); also 'Christianity and 1984' (*Argufying*, pp. 601–4) and the sequence of essays 'Resurrection', 'The Abominable Fancy', 'Heaven and Hell', 'The Satisfaction of the Father', 'The Cult of Unnaturalism', and 'Literary Criticism and the Christian Revival', ibid., pp. 614–37.

63. See Empson, *Seven Types of Ambiguity* (op. cit.), pp. 226–33.

64. Ibid., p. 233.

65. See for instance John Crowe Ransom, 'Mr. Empson's Muddles', *Southern Review*, Vol. 4 (1938–9), pp. 322–39.

66. Empson, *Argufying* (op. cit.), p. 107.

67. See notes 57 and 62, above.

68. Empson, as I have noted, expresses some doubt as to just how far this complex of sentiments might apply to 'actual dogs', that is, to anything observable in their nature, disposition, or instinctual traits. His interest is more in the way that 'dog' came to function as a vehicle or code-word for various forms of emergent humanist or counter-theological belief. Still it is worth remarking the existence of recent research which suggests a high level of canine sensitivity to human purposes and intentions. For a review of this work see Vicki Hearne, 'The Cognitive Dog', *New Scientist*, 12 March 1987, pp. 38–40.

69. See for instance R. Richards, *Darwin and the Emergence of Evolutionary Theories of Mind and Behavior* (Chicago: University of Chicago Press, 1987) and J. Maynard Smith, *The Theory of Evolution* (Harmondsworth: Penguin, 1977).
70. Schiffer, *Remnants of Meaning* (op. cit.).
71. See note 23, above.
72. Empson, 'The Royal Beasts' and Other Works, John Haffenden (ed.) (London: Chatto and Windus, 1988).
73. See Quine, *Word and Object* (op. cit.); also 'Indeterminacy of Translation Again', *Journal of Philosophy*, Vol. 84 (1987), pp. 5–10 and entries under note 15, above.
74. On this topic see also Robert Barrett and Roger Gibson (eds), *Perspectives on Quine* (Oxford: Blackwell, 1989); Christopher Hookway, *Quine: Language, experience and reality* (Cambridge: Polity Press, 1987); Robert Kirk, *Translation Determined* (Oxford: Clarendon Press, 1986).
75. Kornblith, *Inductive Inference and Its Natural Ground* and *Naturalized Epistemology* (note 19, above). See also – from a range of philosophical standpoints – Lorraine Code, *Epistemic Responsibility* (Hanover and London: University Press of New England, 1987); Jonathan Dancy, *An Introduction to Contemporary Epistemology* (Oxford: Blackwell, 1985); Alvin Goldman, *Epistemology and Cognition* (Cambridge, Mass.: Harvard University Press, 1986); Dudley Shapere, *Reason and the Search for Knowledge* (Dordrecht: D. Reidel, 1984); Mary Tiles and Jim Tiles, *An Introduction to Historical Epistemology* (Oxford: Blackwell, 1993).
76. See Davidson, 'A Nice Derangement of Epitaphs' (op. cit.); also his oddly inconclusive sequence of essays on 'The Structure and Content of Truth', *Journal of Philosophy*, Vol. 87 (1990), pp. 279–328.
77. On these difficulties with Quine's position see Kornblith, *Inductive Inference* (op. cit.).
78. See also Papineau, *Reality and Representation* and *Philosophical Naturalism* (note 28, above).
79. See notes 23–5, above.
80. Kornblith, *Inductive Inference* (op. cit.), p. 41.
81. Davidson, *Inquiries into Truth and Interpretation* (op. cit.).
82. Papineau, *Reality and Representation* (op. cit.), p. 35.

5. Realism, Truth and Counterfactual Possibility

1. See J.S. Bell, *Speakable and Unspeakable in Quantum Mechanics: Collected papers on quantum philosophy* (Cambridge: Cambridge University Press, 1987); also James T. Cushing and Ernan McMullan (eds), *Philosophical Consequences of Quantum Theory: Reflections on Bell's theorem* (Indiana: University of Notre Dame Press, 1989); Peter Forrest, *Quantum Metaphysics* (Oxford: Blackwell, 1988); Tim Maudlin, *Quantum Non-Locality and Relativity: Metaphysical intimations of modern science* (Oxford: Blackwell, 1993); Michael Redhead, *Incompleteness, Nonlocality and Realism: A prolegomenon to the philosophy of quantum mechanics* (Oxford: Clarendon, 1987).

2. W.V. Quine, 'Two Dogmas of Empiricism', in *From a Logical Point of View* (Cambridge, Mass.: Harvard University Press, 1953), pp. 20–46; see also Quine, *Ontological Relativity and Other Essays* (New York: Columbia University Press, 1969).

3. See for instance Stephen Schiffer, *Remnants of Meaning* (Cambridge, Mass.: University of Massachusetts Press, 1989); also Steven Stich, *The Fragmentation of Reason* (Cambridge, Mass.: MIT Press, 1990).

4. See especially Richard Rorty, *Consequences of Pragmatism* (Brighton: Harvester, 1982); *Contingency, Irony, and Solidarity* (Cambridge: Cambridge University Press, 1989); and *Objectivity, Relativism, and Truth* (Cambridge UP, 1991).

5. See references to Rorty (note 4, above); also – from a range of philosophical standpoints – Donald Davidson, 'The Structure and Content of Truth', *Journal of Philosophy*, Vol. 87 (1990), pp. 279–328; Michael Devitt, *Realism and Truth* (2nd edn, Oxford: Blackwell, 1986); Paul Horwich, *Truth* (Oxford: Blackwell, 1990); Lawrence E. Johnson, *Focusing on Truth* (London: Routledge, 1982); Richard L. Kirkham, *Theories of Truth* (Cambridge, Mass.: MIT Press, 1992); Gerald Vision, *Modern Anti-Realism and Manufactured Truth* (London: Routledge, 1988); Mark Platts (ed.), *Reference, Truth and Reality: Essays on the philosophy of language* (London: Routledge & Kegan Paul, 1980).

6. See especially Schiffer, *Remnants of Meaning* (op. cit.).

7. Rorty, 'Introduction: pragmatism and philosophy', in *Consequences of Pragmatism* (op. cit.), pp. xiii – xlvii. On the causal theory of reference, see Saul Kripke, *Naming and Necessity* (Oxford: Blackwell, 1980) and Stephen P. Schwartz (ed.), *Naming, Necessity, and Natural Kinds* (Ithaca, NY: Cornell University Press, 1977); also David Wiggins, *Sameness and Substance* (Blackwell, 1980).

8. For discussion of these and related issues see Harold Brown, *Perception, Theory, and Commitment: The new philosophy of science* (Chicago: University of Chicago Press, 1977); Hartry Field, 'Theory Change and the Indeterminacy of Reference', *Journal of Philosophy*, Vol. 70 (1973), pp. 462–81 and 'Quine and the Correspondence Theory', *Philosophical Review*, Vol. 83 (1974), pp. 200–28; Michael E. Levin, 'On Theory-Change and Meaning-Change', *Philosophy of Science*, Vol. 46 (1979), pp. 407–24; Ernest Nagel, Sylvain Bromberger and Adolf Grünbaum, *Observation and Theory in Science* (Baltimore: Johns Hopkins University Press, 1971); David Papineau, *Theory and Meaning* (London: Oxford University Press, 1979) and *Reality and Representation* (Oxford: Blackwell, 1987).

9. See works by Kripke and Putnam cited in note 7 (above); also Platts (ed.), *Reference, Truth and Reality* (op. cit.); J.T. Cushing, C.F. Delaney and G. Gutting (eds), *Science and Reality* (Indiana: University of Notre Dame Press, 1984); Jarrett Leplin (ed.), *Scientific Realism* (Berkeley and Los Angeles: University of California Press, 1984); Peter Lipton, *Inference to the Best Explanation* (London: Routledge, 1993); Nicholas Rescher, *Scientific Realism: A reappraisal* (Dordrecht: D. Reidel, 1987); Peter J. Smith, *Realism and the Progress of Science* (Cambridge: Cambridge University Press, 1981).

10. For some striking examples, see Rom Harré, *Great Scientific Experiments* (London: Oxford University Press, 1983); also Rom Harré, *The Philosophies of Science* (Oxford University Press, 1972); Rom Harré and E.H. Madden, *Causal Powers* (Oxford: Blackwell, 1975).

11. See note 7, above.

12. Hilary Putnam, 'Is Semantics Possible?' and 'Meaning and Reference', in Schwartz (ed.), *Naming, Necessity, and Natural Kinds* (op. cit.), pp. 102–32.

13. See especially Wesley C. Salmon, *Scientific Explanation and the Causal Structure of the World* (Princeton, NJ: Princeton University Press, 1984); Roy Bhaskar, *Scientific Realism and Human Emancipation* (London: Verso, 1986).

14. In James Robert Brown's usefully compact summation: 'Putnam's realist period work is collected in the first two volumes of his *Philosophical Papers* (1975). The anti-realist period started with *Meaning and the Moral Sciences* (1976) and hit full stride in *Reason, Truth, and History* (1981). The third volume of his *Philosophical Papers* (1983) and the more recent collection of essays *Realism with a Human Face* (1990) continue in the same anti-realist vein.' (Brown, *Smoke and Mirrors: How science reflects reality* [London: Routledge, 1994], p. 187.) For further evidence of this marked anti-realist drift see Putnam, *Renewing Philosophy* (Cambridge, Mass.: Harvard University Press, 1992) and *Pragmatism: An open question* (Oxford: Blackwell, 1995); also the essays on his work collected in Peter Clark and Bob Hale (eds), *Reading Putnam* (Oxford: Blackwell, 1993).

15. Brown, *Smoke and Mirrors* (op. cit.); *The Laboratory of the Mind: Thought experiments in the natural sciences* (London: Routledge, 1991).

16. Brown, *Smoke and Mirrors*, p. 112.

17. Ibid., p. 103.

18. For a clear-headed account of these issues, see Peter Gibbins, *Particles and Paradoxes: The limits of quantum logic* (Cambridge: Cambridge University Press, 1987); also Peter Forrest, *Quantum Metaphysics* (op. cit.); Richard Healey, *The Philosophy of Quantum Mechanics: An interactive interpretation* (Cambridge: Cambridge University Press, 1989); J. Polkinghorne, *The Quantum World* (Harmondsworth: Penguin, 1986); Alastair I.M. Rae, *Quantum Physics: Illusion or reality?* (Cambridge UP, 1986). For Putnam's essays on logical problems in the interpretation of quantum mechanics, see *Philosophical Papers*, Vol. 2 (Cambridge UP, 1979).

19. Karl R. Popper, *Quantum Theory and the Schism in Physics* (London: Hutchinson, 1982). See also Joseph Agassi, 'Duhem *versus* Galileo an instrumentalist counterattack on essentialism' and 'The Future of Berkeley's Instrumentalism: just terrific as a self-observer', in *The Gentle Art of Philosophical Polemics* (La Salle, Illinois: Open Court, 1988), pp. 43–54 & 55–69; Donald Gillies, *Philosophy of Science in the Twentieth Century: four central themes* (Oxford: Blackwell, 1993).

20. Brown, *Smoke and Mirrors* (op. cit.), p. 92.

21.   Ibid., p. 93. On this topic of so-far uninstantiated laws or universals, see D.M. Armstrong, *What Is a Law of Nature?* (Cambridge: Cambridge University Press, 1983).

22.   See for instance Niels Bohr, *Atomic Theory and the Description of Nature* (Cambridge: Cambridge University Press, 1934); Nancy Cartwright, *How the Laws of Physics Lie* (London: Oxford University Press, 1983); Michael Gardner, 'Realism and Instrumentalism in Nineteenth-Century Atomism', *Philosophy of Science*, Vol. 46, No. 1 (1979), pp. 1–34; Gibbins, *Particles and Paradoxes* (op. cit.); A. Sudbury, *Quantum Mechanics and the Particles of Nature* (New York: Wiley, 1986).

23.   Thomas S. Kuhn, *The Structure of Scientific Revolutions* (2nd edn, revised, Chicago: University of Chicago Press, 1970). See also references to Quine, 'Two Dogmas of Empiricism' and *Ontological Relativity*, note 2, above.

24.   See entries under note 8, above.

25.   Brown, *Smoke and Mirrors* (op. cit.), p. 81.

26.   Hilary Putnam, *Reason, Truth, and History* (Cambridge: Cambridge University Press, 1981), p. 64.

27.   Putnam, *Philosophical Papers*, Volume One (Cambridge: Cambridge University Press, 1975), p. 73.

28.   For a classic example of this strategy, see the opening paragraphs of Rorty's 'Philosophy as a Kind of Writing: An Essay on Derrida', in *Consequences of Pragmatism* (op. cit.), pp. 90–109.

29.   Rorty, *Objectivity, Relativism, and Truth* (op. cit.), p. 39; cited by Brown, p. 30.

30.   See for instance C.L. Stevenson, *Ethics and Language* (New Haven, Conn.: Yale University Press, 1944) and I.A. Richards, *Principles of Literary Criticism* (London: Paul Trench Trubner, 1924).

31.   Such thinking has various sources and analogues, among them Hans Vaihinger's revisionist reading of Kant (*The Philosophy of As If*), the pragmatism of James and Dewey, and of course Richard Rorty's numerous deconstructions of truth-talk in whatever philosophical guise. A more moderate but technically sophisticated stance is that adopted by Simon Blackburn (*Essays in Quasi-Realism*, London: Oxford University Press, 1993).

32.   See note 27, above.

33.   Hilary Putnam, *Realism with a Human Face* (op. cit.), p. 121.

34.   Ibid., p. 121.

35.   Brown, *Smoke and Mirrors* (op. cit.), p. 86.

36.   See Rorty, *Consequences of Pragmatism* and *Objectivity, Relativism, and Truth* (cited in note 4, above).

37.   Brown, *Smoke and Mirrors*, pp. 85–6.

38.   David Hume, *Enquiries Concerning Human Understanding and Concerning the Principles of Morals*, L.A. Selby-Bigge (ed.) (Oxford: Clarendon Press, 1902), Section VII, Part II, esp. p. 76. For an alternative, more robustly commonsense interpretation of Hume on causality, see Galen Strawson, *The Secret Connexion* (Oxford: Clarendon Press, 1989) and John Wright, *The Sceptical Realism of David Hume*

(Manchester: Manchester University Press, 1983); also Simon Blackburn's dissenting response in his *Essays in Quasi-Realism* (op. cit.).

39. See especially Adolf Grünbaum and Wesley C. Salmon (eds), *The Limitations of Deductivism* (Berkeley and Los Angeles: University of California Press, 1988); Salmon, *Four Decades of Scientific Explanation* (Minneapolis: University of Minnesota Press, 1989).

40. Brown, *Smoke and Mirrors*, p. 97.

41. On thought-experiments in science see Brown, *The Laboratory of the Mind* (op. cit.); also Arthur Fine, *The Shaky Game: Einstein, realism, and quantum theory* (Chicago: University of Chicago Press, 1986); Roy Sorensen, *Thought Experiments* (New York and London: Oxford University Press, 1992); Paul Davies, 'The Thought that Counts: thought experiments in physics', *New Scientist*, Vol. 146 (May 6 1995), pp. 26–31.

42. See Quine, 'Two Dogmas of Empiricism' (op. cit.).

43. For further discussion of these and related issues, see M.J. Loux (ed.), *The Possible and the Actual: Readings in the metaphysics of modality* (Ithaca, NY: Cornell University Press, 1979); J.S. Bell, 'Six Possible Worlds in Quantum Mechanics', in Sture Allen (ed.), *Possible Worlds in the Humanities, Arts, and Sciences* (Berlin and New York: Walter de Gruyter, 1989), pp. 359–73; David Lewis, *Counterfactuals* (Cambridge, Mass.: Harvard University Press, 1973) and *On the Plurality of Worlds* (Oxford: Blackwell, 1986); Alvin Plantinga, *The Nature of Necessity* (Oxford: Clarendon Press, 1974); Ruth Ronen, *Possible Worlds in Literary Theory* (Cambridge: Cambridge University Press, 1994).

44. Brown, *Smoke and Mirrors* (op. cit.), p. 112.

45. See for instance Nelson Goodman's various 'new puzzles' of induction as propounded in his book *Fact, Fiction and Forecast* (Cambridge, Mass.: Harvard University Press, 1955); also – from a more moderate anti-realist standpoint – Cartwright, *How the Laws of Nature Lie* (op. cit.) and Bas van Frassen, *The Scientific Image* (London: Oxford University Press, 1980) and *Laws and Symmetries* (Oxford University Press, 1989).

46. Nelson Goodman, *Ways of Worldmaking* (Indianapolis: Bobbs-Merrill, 1978).

47. Goodman, *Fact, Fiction and Forecast* (op. cit.).

48. Goodman, *Ways of Worldmaking* (op. cit.).

49. See Quine, 'Goodman's *Ways of Worldmaking*', in *Theories and Things* (Cambridge, Mass.: Harvard University Press, 1981); also Joseph Margolis, *Texts without Referents* (Oxford: Blackwell, 1989) and *The Truth About Relativism* (Blackwell, 1991), especially pp. 113–17.

50. Nelson Goodman, *Of Mind and Other Matters* (Cambridge, Mass.: Harvard University Press, 1984), p. 127.

51. Ibid., p. 127.

52. Ibid., p. 38.

53. Margolis, *Texts without Referents* (op. cit.), p. 110.

54. See especially Margolis, *Science without Unity* (Oxford: Blackwell, 1991).

55. Goodman, *Of Mind and Other Matters* (op. cit.), p. 38.
56. Ibid., p. 38.
57. Margolis, *Texts without Referents* (op. cit.), p. 110.
58. Margolis, *Pragmatism without Foundations* (Oxford: Blackwell, 1986); *Science without Unity* (op. cit.); and *Texts without Referents: Reconciling science and narrative* (Blackwell, 1987). See also Margolis, *The Truth About Relativism* (op. cit.).
59. See for instance Steve Fuller, *Philosophy of Science and Its Discontents* (Boulder, Colorado: Westview Press, 1989); Joseph Rouse, *Knowledge and Power: Toward a political philosophy of science* (Ithaca, NY: Cornell University Press, 1987); also – from a strongly dissenting viewpoint – Paisley Livingston, *Literary Knowledge: Humanistic inquiry and the philosophy of science* (Ithaca, NY: Cornell University Press, 1988).
60. Hans-Georg Gadamer, *Truth and Method*, trans. Garrett Barden and John Cumming (New York: Seabury Press, 1975), p. 271.
61. Margolis, *Texts without Referents* (op. cit.), p. 309.
62. Ibid., p. 310.
63. See for instance Richard Rorty, *Essays on Heidegger and Others* (Cambridge: Cambridge University Press, 1991); also Mark Okrent, *Heidegger's Pragmatism* (Ithaca, NY: Cornell University Press, 1988) and Hubert L. Dreyfus, *Being-in-the-World: A commentary on Heidegger's Being and Time, Division 1* (Cambridge, Mass.: MIT Press, 1991).
64. Bertrand Russell, *The Problems of Philosophy* (London: Oxford University Press, 1912), pp. 129–30.

6.   Stones and Pendulums

1. See especially Joseph Margolis, *The Truth about Relativism* (Oxford: Blackwell, 1991) and *Interpretation Radical but Not Unruly: The new puzzle of the arts and history* (Berkeley and Los Angeles: University of California Press, 1995).
2. For some representative arguments see Martin Hollis and Steven Lukes (eds), *Rationality and Relativism* (Oxford: Blackwell, 1982); also Robert L. Arrington, *Rationalism, Realism, and Relativism: Perspectives in contemporary moral epistemology* (Ithaca, NY: Cornell University Press, 1989); Jeffrey Alexander, *Fin de Siècle Social Theory: Relativism, reduction and the problem of reason* (London: Verso, 1995); Richard J. Bernstein, *Beyond Objectivism and Relativism: science, hermeneutics, and praxis* (Philadelphia, PA.: University of Pennsylvania Press, 1983); Larry Laudan, *Science and Relativism: Some key issues in the philosophy of science* (Chicago: University of Chicago Press, 1990); W.H. Newton-Smith, *The Rationality of Science* (London: Routledge & Kegan Paul, 1981).
3. Nelson Goodman, *Of Mind and Other Matters* (Cambridge, Mass.: Harvard University Press, 1984), p. 127.
4. See Margolis, *Texts without Referents: Reconciling science and narrative* (Oxford: Blackwell, 1987), especially pp. 108–11.
5. W.V. Quine, 'Goodman's *Ways of Worldmaking*', in *Theories and Things* (Cambridge, Mass.: Harvard University Press, 1981).

6. Margolis, *The Truth about Relativism* (op. cit.).
7. Margolis, *Texts without Referents* (op. cit.); see also his earlier volumes in the sequence, *Pragmatism without Foundations* (Oxford: Blackwell, 1986) and *Science without Unity* (Blackwell, 1987).
8. Margolis, *Texts without Referents* (op. cit.), especially pp. 293–315.
9. Arthur C. Danto, *Narration and Knowledge* (New York: Columbia University Press, 1985), p. 67.
10. See for instance R.G. Collingwood, *The Idea of History* (Oxford: Clarendon Press, 1946); Hans-Georg Gadamer, *Truth and Method*, trans. and ed. Garrett Barden and John Cumming (London: Sheed & Ward, 1975); Paul Ricoeur, *Hermeneutics and the Human Sciences: Essays on language, action, and interpretation*, trans. and ed. John B. Thompson (Cambridge: Cambridge University Press, 1981).
11. Danto, *Narration and Knowledge* (op. cit.), pp. 326–7.
12. Ibid., p. 330.
13. See for instance the essays and interviews collected in Michel Foucault, *Language, Counter-Memory, Practice*, D.F. Bouchard and S. Weber (eds) (Oxford: Blackwell, 1977).
14. Danto, *Narration and Knowledge*, pp. 336–7.
15. Thomas S. Kuhn, *The Structure of Scientific Revolutions*, 2nd edn, revised (Chicago: University of Chicago Press, 1970); also Gary Gutting, *Paradigms and Revolutions* (Notre Dame, Ind.: Notre Dame University Press, 1980); Ian Hacking (ed.), *Scientific Revolutions* (London: Oxford University Press, 1981); Paul Horwich (ed.), *The World Changes: Thomas Kuhn and the nature of science* (Cambridge, Mass.: MIT Press, 1993); John Krige, *Science, Revolution and Discontinuity* (Brighton: Harvester, 1980).
16. Cited by Margolis, *Texts without Referents* (op. cit.), p. 298.
17. See Kuhn's 1969 Postscript to *The Structure of Scientific Revolutions* (op. cit.); also Kuhn, *The Essential Tension: Selected studies in scientific tradition and change* (Chicago: University of Chicago Press, 1977) and 'Rationality and Theory-Choice', *Journal of Philosophy*, Vol. 80 (1983), pp. 563–70.
18. Danto, *Narration and Knowledge*, p. 337.
19. See for instance Hilary Putnam, *The Many Faces of Realism* (La Salle: Open Court, 1987) and *Realism with a Human Face* (Cambridge, Mass.: Harvard University Press, 1990).
20. See especially Adolf Grünbaum and Wesley C. Salmon (eds), *The Limitations of Deductivism* (Berkeley & Los Angeles: University of California Press, 1988); also Salmon, *The Foundations of Scientific Inference* (Pittsburgh: University of Pittsburgh Press, 1967) and *Four Decades of Scientific Explanation* (Minneapolis: University of Minnesota Press, 1989).
21. Aristotle, *De Interpretatione*, trans. J.L. Ackrill (London: Oxford University Press, 1963). See also Michael Dummett, 'Can an Effect Precede Its Cause?', 'Bringing about the Past', and 'The Reality of the Past', in *Truth and Other Enigmas* (London: Duckworth, 1978), pp. 319–32, 333–50 and 358–74.

22. Martin Heidegger, *Being and Time*, trans. J. Macquarrie and E. Robinson (Oxford: Blackwell, 1962); also Heidegger, 'On the Essence of Truth', trans. John Sallis, in David F. Krell (ed.) *Martin Heidegger: Basic Writings*, (London: Routledge & Kegan Paul, 1982), pp. 117–41.

23. T.W. Adorno, *The Jargon of Authenticity*, trans. K. Tarnowski and F. Will (London: Routledge & Kegan Paul, 1973).

24. See for instance Hubert M. Dreyfus, *Being-in-the-World: A commentary on Heidegger's Being and Time, Division I* (Cambridge, Mass.: MIT Press, 1991); J.E. Malpas, *Donald Davidson and the Mirror of Meaning* (Cambridge: Cambridge University Press, 1992); Mark Okrent, *Heidegger's Pragmatism* (Ithaca, NY: Cornell University Press, 1988).

25. Christopher Norris, 'Settling Accounts: Heidegger, de Man, and the ends of philosophy', in *What's Wrong with Postmodernism* (Hemel Hempstead: Harvester-Wheatsheaf, 1990), pp. 222–83; also Norris, 'Getting at Truth: Genealogy, Critique, and Postmodern Scepticism', in *The Truth about Postmodernism* (Oxford: Blackwell, 1993), pp. 257–304.

26. Heidegger, *Being and Time* (op. cit.), p. 195.

27. See works cited in note 24, above.

28. See especially Richard Rorty, 'Overcoming the Tradition: Heidegger and Dewey', in *Consequences of Pragmatism* (Brighton: Harvester, 1982), pp. 37–59; also Rorty, *Essays on Heidegger and Others* (Cambridge: Cambridge University Press, 1991).

29. See Margolis, *Science without Unity* (op. cit.).

30. See Danto, *The Transfiguration of the Commonplace* (Cambridge, Mass.: Harvard University Press, 1981).

31. See for instance Hayden White, *Tropics of Discourse* (Baltimore: Johns Hopkins University Press, 1978) and *The Content of the Form* (Johns Hopkins U.P., 1987).

32. On this topic see especially Peter Geach, *Reference and Generality* (Ithaca, NY: Cornell University Press, 1962); Nicholas Griffin, *Relative Identity* (Oxford: Clarendon Press, 1977); David Wiggins, *Identity and Spatio-Temporal Continuity* (Oxford: Blackwell, 1967) and *Sameness and Substance* (Blackwell, 1980).

33. See entries under note 32 (above) for a range of views on the usefulness or otherwise of this talk about 'relative identity'.

34. See Rorty, *Consequences of Pragmatism* (op. cit.) and *Contingency, Irony, and Solidarity* (Cambridge: Cambridge University Press, 1989).

35. W.V. Quine, *Word and Object* (Cambridge, Mass.: MIT Press, 1960), p. 221.

36. See especially Wesley C. Salmon, *Scientific Explanation and the Causal Structure of the World* (Princeton, NJ: Princeton University Press, 1984); Rom Harré, *Varieties of Realism: A rationale for the social sciences* (Oxford: Blackwell, 1986) and *The Principles of Scientific Thinking* (Chicago: University of Chicago Press, 1970); Rom Harré and E.H. Madden, *Causal Powers* (Blackwell, 1975); Roy Bhaskar, *Scientific Realism and Human Emancipation* (London: Verso, 1986) and *Reclaiming Reality: A critical introduction to contemporary philosophy* (London: Verso, 1989).

37. Bhaskar, *Scientific Realism and Human Emancipation* (op. cit.).
38. See works cited in note 21, above.
39. See for instance J.S. Bell, *Speakable and Unspeakable in Quantum Mechanics* (Cambridge: Cambridge University Press, 1987); James T. Cushing and Ernan McMullin (eds), *Philosophical Consequences of Quantum Theory: Reflections on Bell's theorem* (Indiana: University of Notre Dame Press, 1989); Tim Maudlin, *Quantum Non-Locality and Relativity: metaphysical intimations of modern science* (Oxford: Blackwell, 1993); Michael Redhead, *Incompleteness, Nonlocality and Realism* (Oxford: Clarendon Press, 1987).
40. Richard Swinburne, *The Christian God* (Oxford: Clarendon Press, 1994), p. 85.
41. Michael Dummett, 'Causal Loops', in R. Flood and M. Lockwood (eds), *The Nature of Time* (Oxford: Blackwell, 1986). See also Dummett, 'Can an Effect Precede Its Cause?', 'Bringing about the Past', and 'The Reality of the Past', cited in note 21, above.
42. Swinburne, *The Christian God* (op. cit.), p. 88.
43. Ibid., p. 88.
44. Ibid., p. 61.
45. Ibid., pp. 61–2.
46. Ibid., p. 157.
47. Ibid., pp. 89–90.
48. See Saul Kripke, *Naming and Necessity* (Oxford: Blackwell, 1980); David Wiggins, *Sameness and Substance* (op. cit.); Stephen Schwartz (ed.), *Naming, Necessity, and Natural Kinds* (Ithaca, NY: Cornell University Press, 1977).
49. Swinburne, op. cit., p. 26.
50. Ibid., p. 31.
51. Nelson Goodman, *Fact, Fiction and Forecast* (Cambridge, Mass.: Harvard University Press, 1955).
52. See Putnam, *The Many Faces of Realism and Realism with a Human Face* (cited in note 19, above).
53. Danto, *Narration and Knowledge* (op. cit.), p. 67.

Index of Names